Series Editor's Foreword

Organometallic chemistry has been one of the most exciting growth areas of inorganic chemistry over the last 40 years. As well as being an established academic discipline it is of key importance in the design of industrial homogeneous catalysts, and also of new organic synthetic methods.

Oxford Chemistry Primers are designed to give a concise introduction to all chemistry students by providing the material that would usually be covered in an 8–10 lecture course. As well as providing up-to-date information, this series will provide explanations and rationales that form the framework of an understanding of inorganic chemistry. In the first of his books, Manfred Bochmann provides the reader with a clear, up-to-date description of the chemistry of single-carbon donor ligands in transition metal complexes, building upon well-accepted fundamentals.

John Evans
Department of Chemistry, University of Southampton

Preface

Organometallic chemistry has rapidly increased in importance over the past few decades and is now a firm part of any chemistry curriculum. It is fundamental to our understanding of catalysis and is becoming increasingly important for the development of new synthetic methods and materials. This volume, the first of two parts, sets out the most important structural principles in what might at first appear a bewildering variety of compounds and introduces some of the major classes of organometallic transition metal complexes: metal carbonyls, metal alkyls, and alkylidenes and alkylidynes. *Organometallics 2* will cover π-complexes. Both booklets are intended to complement current general textbooks on inorganic chemistry which are often unable to cover this area of chemistry adequately.

In each chapter, the most pertinent synthetic methods are illustrated by typical examples, followed by an outline of characteristic reactivity patterns which, it is hoped, will allow students to develop a feel for the synthetic potential of this chemistry. Variety in the text is provided by highlighted sections which either draw attention to a particular general principle of reactivity or structure, or form an excursion into important applications, in particular in homogeneous catalysis and organic synthesis. While references to the primary literature could not be included, brief bibliographies will lead the reader to more comprehensive reviews and reference texts.

Norwich
November 1992

M.B.

Contents

Organometallics 1

Complexes with transition metal–carbon σ-bonds

Manfred Bochmann

University of East Anglia

OXFORD   NEW YORK   TOKYO
OXFORD UNIVERSITY PRESS
1994

Oxford University Press, Walton Street, Oxford OX2 6DP

Oxford New York Toronto
Delhi Bombay Calcutta Madras Karachi
Kuala Lumpur Singapore Hong Kong Tokyo
Nairobi Dar es Salaam Cape Town
Melbourne Auckland Madrid

and associated companies in
Berlin Ibadan

Oxford is a trade mark of Oxford University Press

Published in the United States
by Oxford University Press Inc., New York

A catalogue record for this book is available from the British Library

Library of Congress Cataloging-in-Publication Data
Bochmann, Manfred.
Organometallics 1: complexes with transition metal-carbon σ-bonds/Manfred Bochmann.
(Oxford chemistry primers;)
Includes bibliographical references and index.
1. Organotransition metal compounds. I. Title. II. Series.
QD411.8.T73B63   1993   547'.056—dc20   93-22883
ISBN 0-19-855751-5 (Hbk)
ISBN 0-19-855750-7 (Pbk)

Typeset by Pentacor PLC, High Wycombe
Printed in Great Britain by
Bath Press Ltd, Bath, Avon

1 A few basics

1.1 Introduction

Organometallic compounds are defined as substances containing direct metal–carbon bonds. The variety of the organic moiety in such compounds is practically infinite, ranging from alkyl substituents to alkenes, alkynes, carbonyls, and aromatic and heterocyclic compounds. Although some organometallic compounds have been known for a long time, it is only in the last four or five decades that organometallic chemistry has come into its own and has experienced tremendous growth, both at a fundamental level where our insight into the nature of chemical bonds has been broadened by a surprising variety of bonding situations without precedence elsewhere, and in its economic impact, such as catalysis.

Probably the first organometallic compound was prepared over 200 years ago, when, in 1760, the French chemist L. C. Cadet during attempts to make invisible inks from arsenate salts prepared a repulsively smelling liquid which was later identified as dicacodyl, As_2Me_4 (Greek κακοδια = stink). Dicacodyl is a typical representative of an 'electron-precise' organometallic compound of a main group element: the valence electrons are paired with valence electrons of carbons to make σ bonds, and the element is in a high oxidation state. Similar alkyl compounds $M(CH_2R)_n$ exist for a range of elements, from lithium to tellurium, in which the hydrocarbyl moiety CH_2R has varying degrees of carbonionic character and the polarity of the M–C bonds decreases from highly ionic (M = heavy alkali metal) to more or less covalent (M = p-block element). The element–carbon bonds are formed utilizing available s and p orbitals.

However, it is with the transition metals that the full diversity of organometallic chemistry becomes apparent. In addition to s and p orbitals these elements possess a shell of d orbitals ideally suited for overlap for example with π-orbitals of unsaturated organic molecules: d_{xy}, d_{xz}, d_{yz}, $d_{x^2-y^2}$ and d_{z^2}:

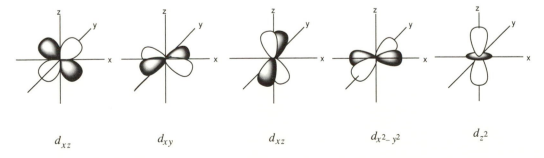

d_{xz} d_{xy} d_{xz} $d_{x^2-y^2}$ d_{z^2}

The metal orbitals can hybridize, leading to a wide range of coordination geometries and coordination numbers (1–8, through 4–6 are more typical), and they can interact with the π and π^* levels of unsaturated organic molecules by donating as well as accepting electron density, thus changing reactivity patterns and facilitating otherwise inaccessible transformations.

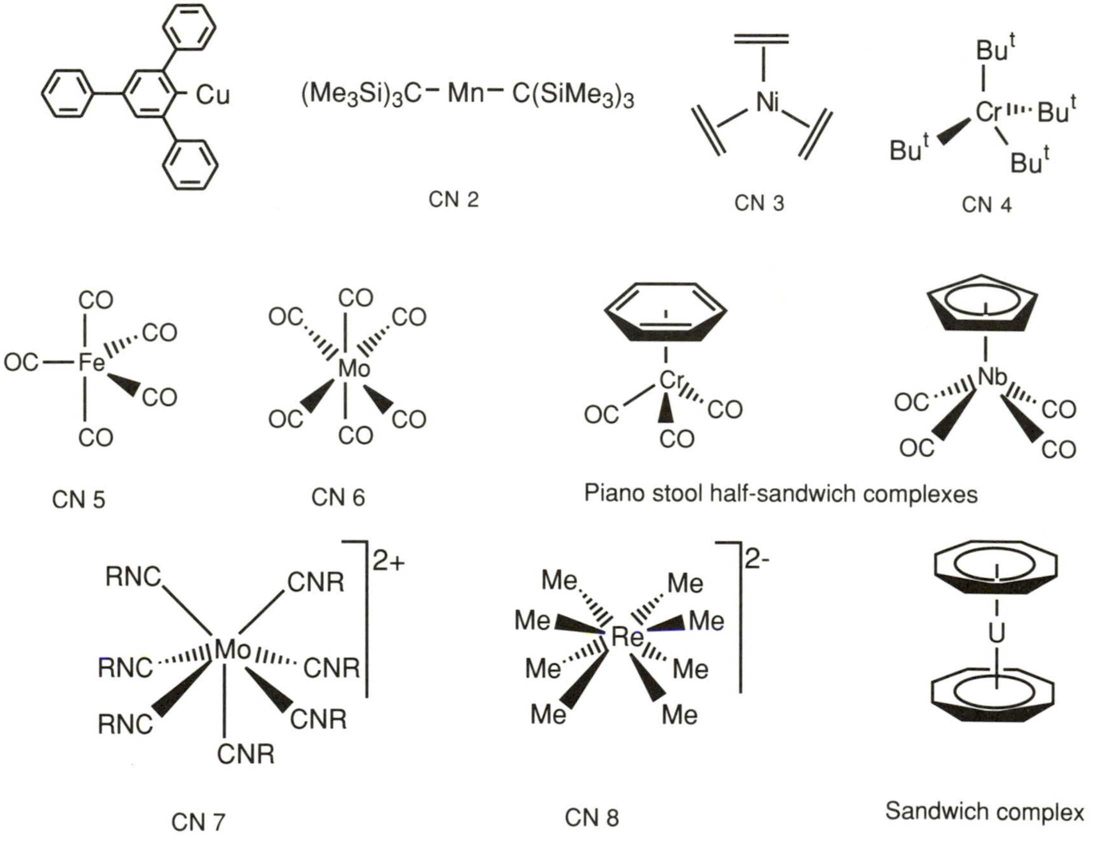

Cu
$(Me_3Si)_3C-Mn-C(SiMe_3)_3$
CN 2
Ni
CN 3
Bu^t
Cr
CN 4
CO
OC
Fe
CN 5
Mo
CN 6
Cr
Nb
Piano stool half-sandwich complexes
RNC
CNR
2+
Mo
CN 7
Me
Re
2-
CN 8
U
Sandwich complex

Examples for coordination geometries of transition and actinide complexes with coordination numbers from 1 to 8.

1.2 The 18-electron rule

A main group element such as tin uses its four valence electrons for the formation of four σ bonds ('2-electron-2-centre' bonds); the result is a stable molecule such as $SnMe_4$. The stability of $SnMe_4$ – and the instability of, for example, $SnMe_3$ – can be predicted by the *octet rule*: a sufficient number of bonds is formed to surround the central element by an octet of electrons: in this way the element reaches the electron configuration of the noble gas with the next highest atomic number.

For transition metals the *18-electron rule* has a similar predictive function. It provides a simple 'rule-of-thumb' basis for the discussion of structure and bonding, and although there are numerous exceptions, leafing through this book will show that compounds which fulfil this rule tend to exhibit particular stability.

> **The *18-electron rule*:** A stable complex (with the electron configuration of the next highest noble gas) is obtained when the sum of metal *d*-electrons, electrons donated from the ligands, and of the overall charge of the complex equals 18.

Since 18 electrons are required to fill the five *d*-, one *s*- and three *p*-orbitals of a transition metal, the electron configuration of the next highest noble gas is once again reached. For example, zerovalent nickel in $Ni(CO)_4$ has a completely filled shell of 10 *d*-electrons, and since each CO binds through a lone electron pair on carbon, the ligands contribute $4 \times 2 = 8$ electrons: $d^{10} + 8 = 18$. Similarly, the count for $Fe(CO)_5$ is $d^8 + 5 \times 2 = 18$.

> NOTE: Although textbooks will frequently give the electron configuration of a zerovalent transition metal such as Ti as $[Ar]3d^2 4s^2$ and of Fe as $[Ar]3d^6 4s^2$, in a chemical environment the 4*s* level is *always* higher in energy than 3*d*: Ti(0) has the configuration $[Ar]3d^4$, Fe is $[Ar]3d^8$ (F. L. Pilar, 1978).

4*s* is ALWAYS above 3*d*!

Electron counting conventions: it is useful to treat anionic (or cationic) ligands always as neutral radicals and the metal centre as zerovalent – this avoids discussions about formal oxidation states. The electron count is of course identical to the alternative approach of distributing charges 'realistically' and treating metals as cations and ligands as anions:

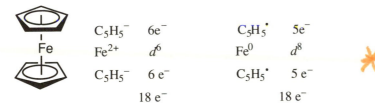

	$C_5H_5^-$	$6e^-$		$C_5H_5^{\bullet}$	$5e^-$
Fe	Fe^{2+}	d^6		Fe^0	d^8
	$C_5H_5^-$	$6 e^-$		$C_5H_5^{\bullet}$	$5 e^-$
		$18 e^-$			$18 e^-$

Ionic counting convention   Covalent convention

A word about formal oxidation states. Since $Na^+C_5H_5^-$ exists as an ionic substance, it is reasonable to consider C_5H_5 in ferrocene as anionic as well, similar to Cl^-, and attribute a formal oxidation state of +II to iron. However, it must be remembered that bonding in complexes such as ferrocene is highly covalent. Calculations suggest that Fe accumulates only a charge of +0.2 and each C_5H_5 ligand –0.1: the covalent model is just as valid. With other ligands the situation is less straightforward: is η^7-C_7H_7 (cycloheptatrienyl — see *Organometallics 2*, Chapter 6) for example to be considered as $C_7H_7^+$ (6π Hückel aromatic), $C_7H_7^{\bullet}$ (radical), $C_7H_7^-$ (8π antiaromatic), or even as $C_7H_7^{3-}$ (10π aromatic)?

Electron counting and formal oxidation states.

Odd electron complex fragments can acquire an additional electron by dimerizing and forming a metal–metal bond:

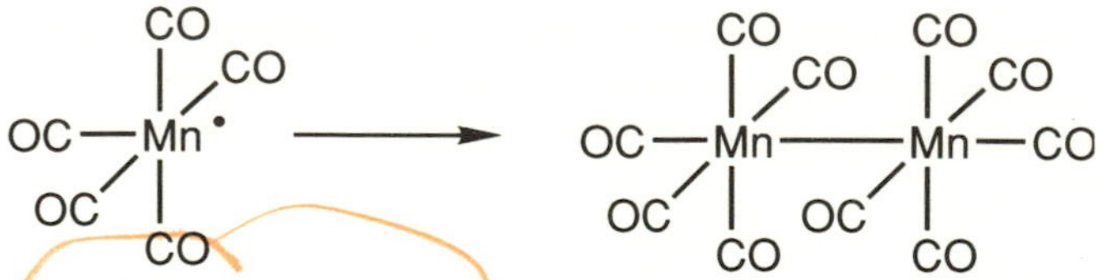

$d^7 + 5 \times 2 = 17$ VE $d^7 + 5 \times 2 + 1 = 18$ VE

σ-Bonded ligands may utilize lone electron pairs in additional 2-electron donor interactions:

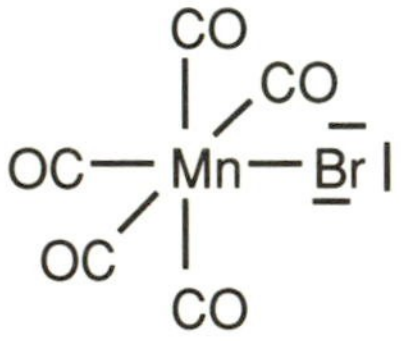

Br as 1-electron ligand:

Mn^0	7 e$^-$
5 COs	10 e$^-$
Br	1 e$^-$
	18 VE

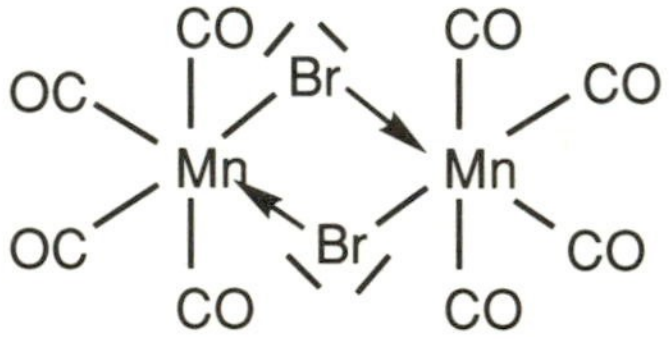

Br as 3-electron ligand:

Mn^0	7 e$^-$
4 COs	8 e$^-$
1 Br	1 e$^-$
1 Br	2 e$^-$
	18 VE

The vast majority of organometallic transition metal complexes obey the 18-electron rule. Notable exceptions are electron-rich complexes of the late transition metals with d^8 configuration, such as Rh(I), Ir(I), Ni(II), Pd(II), and Pt(II): these have a strong preference to form square-planar 16-electron complexes. The reason is the increased stabilization of the d-shell as the atomic number increases, so that, for example, the occupied d_{z^2} orbital no longer participates in ligand bonding, and stabilization through back-bonding (see bonding of CO in carbonyl complexes) becomes less efficient. Characteristic types of d^8 complexes are, for example:

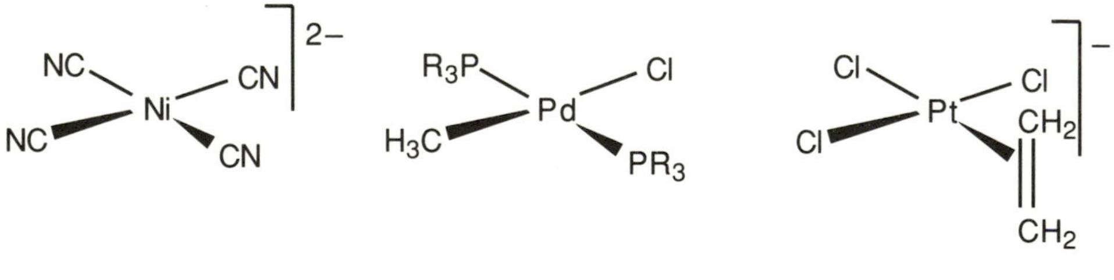

For electron counting purposes ligands can be classified according to the number of electrons they formally donate to the metal centre, following the covalent (electroneutral) counting convention:

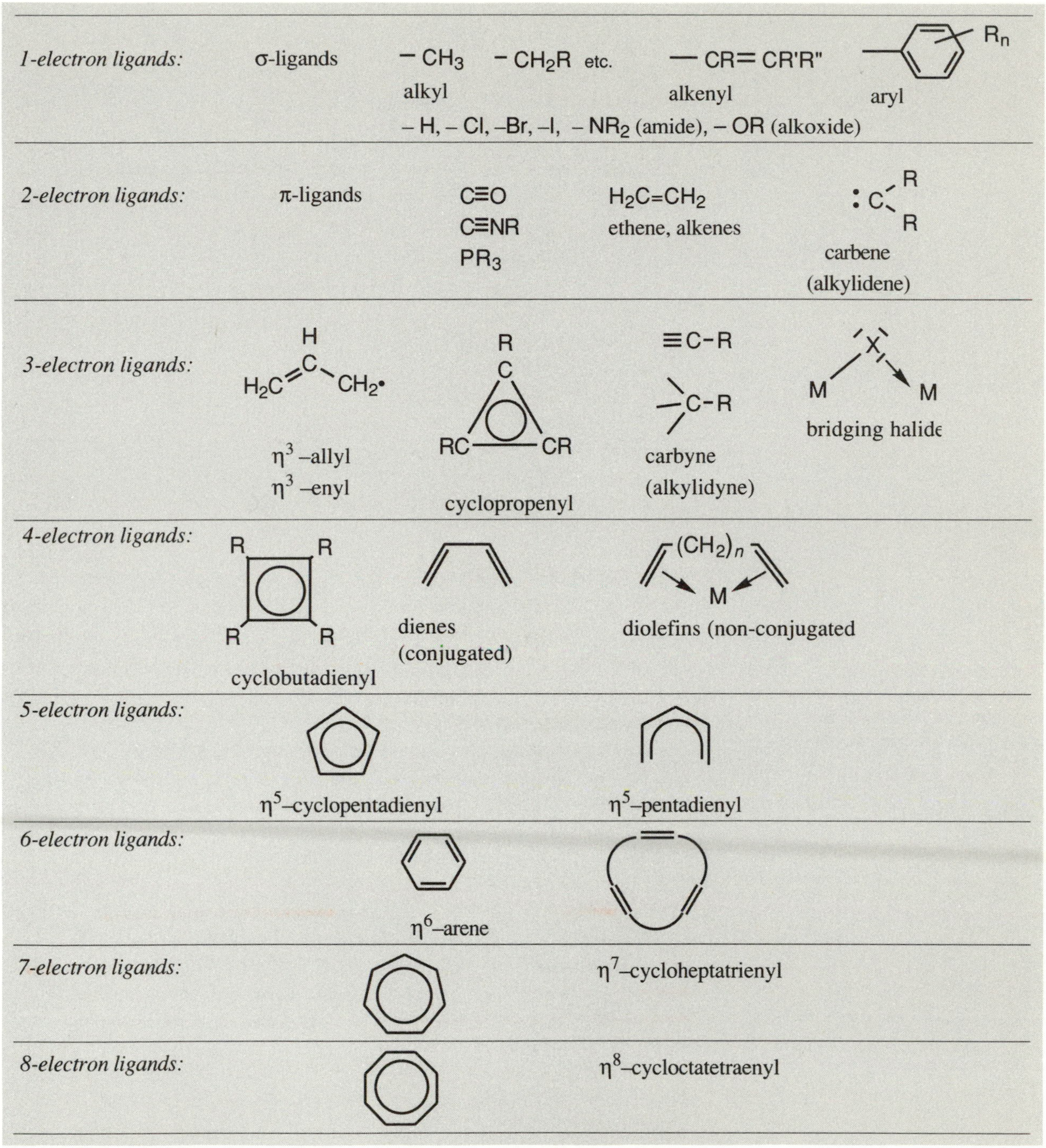
1-electron ligands:
σ-ligands
– CH3
– CH2R etc.
alkyl
— CR= CR'R"
alkenyl
Rn
aryl
– H, – Cl, –Br, –I, – NR2 (amide), – OR (alkoxide)
2-electron ligands:
π-ligands
C≡O
C≡NR
PR3
H2C=CH2
ethene, alkenes
R
R
carbene
(alkylidene)
3-electron ligands:
H
H2C
C
CH2•
η3 –allyl
η3 –enyl
R
C
RC
CR
cyclopropenyl
≡C-R
C-R
carbyne
(alkylidyne)
X
M
M
bridging halide
4-electron ligands:
R
R
R
R
cyclobutadienyl
dienes
(conjugated)
(CH2)n
M
diolefins (non-conjugated
5-electron ligands:
η5–cyclopentadienyl
η5–pentadienyl
6-electron ligands:
η6–arene
7-electron ligands:
η7–cycloheptatrienyl
8-electron ligands:
η8–cycloctatetraenyl

With unsaturated ligands it is necessary to specify the number of carbon atoms which interact with the metal centre. The prefix η^n before the ligand formula implies bonding to n carbons, while μ_k indicates a ligand bridging k metal atoms. Individual numbers of ligand atoms may be required to describe more complicated structures:

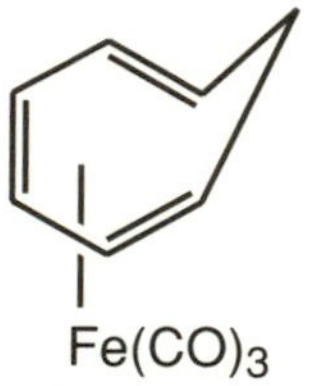

$(\eta^4\text{-}C_7H_8)Fe(CO)_3$

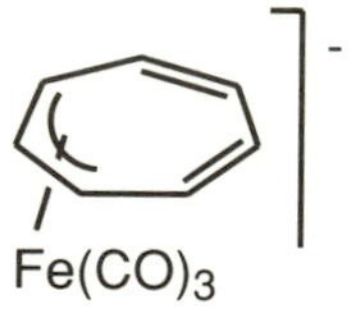

$(\eta^3\text{-}C_7H_7)Fe(CO)_3$

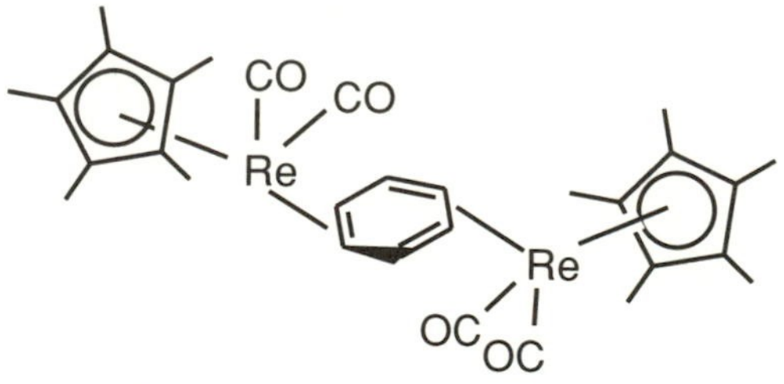

$[\{(\eta^5\text{-}C_5Me_5)Re(CO)_2\}_2\{\mu_2\text{-}(\eta^2{:}\eta^2\text{-}C_6H_6)\}]$

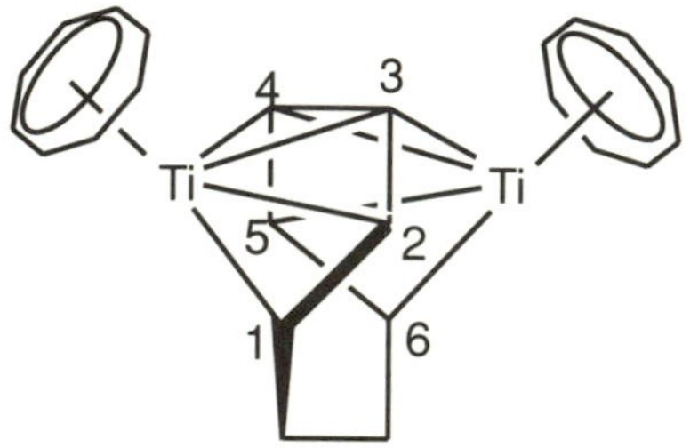

$[(\eta^8\text{-}C_8H_8)Ti]_2[\mu\text{-}(1\text{-}4\eta{:}3\text{-}6\eta\text{-}C_8H_8)$

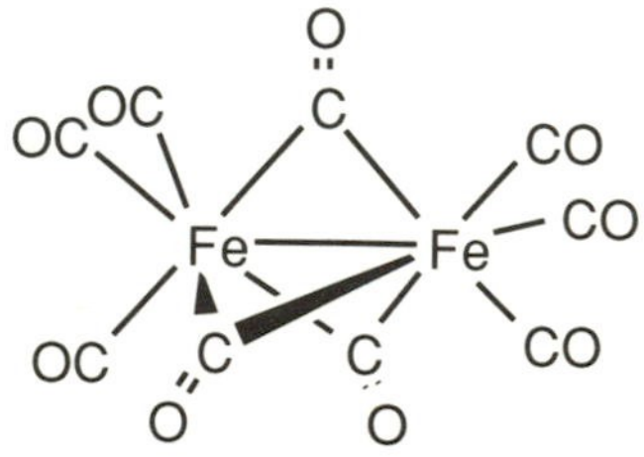

$Fe_2(\mu_2\text{-}CO)_3(CO)_6$

Why do classical coordination complexes often not obey the 18-electron rule, and how do organometallic compounds relate to them?

Whereas the 18-electron rule is obeyed by the majority of organometallic complexes, many classical coordination compounds, such as ammine and aquo complexes, have widely differing electron counts and frequently exceed the magic number 18. Why?

For an octahedral complex with six σ-donor ligands, the molecular orbital diagram shows that the metal d-levels are split into more stable triplet (t_{2g}) and less stable doublet (e_g) states by an energy difference Δ.

Three classes of complexes can be distinguished:

CLASS I: Δ small: n VE $\geq$ or $\leq$ 18, M = first-row transition metal.
For first-row transition metals coordinated to ligands inducing a weak ligand field, such as F^-, Cl^-, H_2O, or NH_3, the ligand field splitted Δ is small; t_{2g} is essentially non-bonding and e_g only weakly antibonding. Both t_{2g} and e_g can be occupied with up to 10 electrons, for example:

		VE
TiF_6^{2-}	Ti^{IV}, d^0	12
$Mn(CN)_6^{3-}$	Mn^{III}, d^4	16
$Co(NH_3)_6^{2+}$	Co^{II}, d^7	19
$Zn(en)_3^{2+}$	Zn^{II}, d^{10}	22

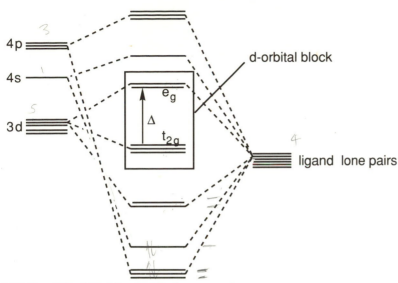

CLASS II: *n* VE ≤ 18, M = 2nd/3rd row metal.

For second- or third-row transition metals and ligands in the upper ranges of the spectrochemical series, t_{2g} is non-bonding and occupied with 0–6 electrons, e_g is antibonding and unoccupied.

		VE
WCl_6	W^{VI}, d^0	12
WCl_6^{2-}	W^{IV}, d^2	14
$W(CN)_8^{3-}$	W^V, d^1	17
PtF_6^{2-}	Pt^{IV}, d^6	18
$PtCl_4^{2-}$	Pt^{II}, d^8	16

CLASS III: Δ large: *n* VE = 18; organometallic complexes.

Most ligands in this class are good electron acceptors as well as donors and form strong bonds to the metal centre. Δ is large, t_{2g} becomes bonding and is fully occupied, e_g is antibonding and empty. The 18-electron rule is obeyed as far as steric constraints will allow.

		VE
$Ti(CO)_5(dmpe)$	Ti^0, d^4	18
$V(CO)_6^-$	V^{-I}, d^6	18
$Fe(CO)_5$	Fe^0, d^8	18
$CpCo(CO)_2$	Co^0, d^9	18
$Ni(COD)_2$	Ni^0, d^{10}	18

Exceptions (on steric grounds) include $Ti(CH_2SiMe_3)_4$ (8 VE), $ReMe_6$ (13 VE), Cp_2ZrCl_2 (16 VE), and $V(CO)_6$ (17 VE).

2 Metal carbonyl complexes

In 1888 Ludwig Mond, one of the founders of what was later to become ICI (Imperial Chemical Industries), observed during experiments with carbon monoxide that an unexpected volatile substance was formed when CO had been in contact with nickel. This substance was isolated as a heavy, almost colourless liquid which contained nickel but possessed properties unlike any nickel compound known so far. It boiled at only 34°C and decomposed on heating cleanly to metallic nickel with the release of four CO molecules per nickel atom: a complex between CO and nickel had formed of the composition $Ni(CO)_4$. Within a few years the practically-minded Mond had turned his observation into a large scale process for the manufacture of high purity nickel; his plant in South Wales is still in operation and one of the largest of its kind. Shortly afterwards a second carbonyl complex, $Fe(CO)_5$, was prepared, independently by Mond and by Berthelot in France, from CO and finely divided iron. The carbonyls of other metals are best prepared from metal salts and CO under reducing conditions.

The physical properties of these volatile metal carbonyls were so unusual that they quickly caught the imagination of chemists worldwide, and metal carbonyl complexes have remained one of the most important classes of organometallic compounds ever since.

CO binds firmly to blood haemoglobin and renders it unavailable for oxygen transport. Metal carbonyls, in particular the volatile compounds, are therefore highly toxic; $Ni(CO)_4$ is a very potent carcinogen.

2.1 Neutral metal carbonyls

Synthesis
From the metal

$$Ni + 4\,CO \xrightarrow{\text{1 bar, 25°C}} Ni(CO)_4$$
Colourless liquid, b.p. 34°C

$$Fe + 5\,CO \xrightarrow{\text{100 bar, 150°C}} Fe(CO)_5$$
Yellow liquid, b.p. 103°C

From metal salts

$$TiCl_4 + 6\ KC_{10}H_8 \xrightarrow[\text{15-crown-5}]{\text{1 bar CO}} [K(15\text{-crown-5})^+]_2[Ti(CO)_6]^{2-} + 4\ KCl + 6\ C_{10}H_8$$

Note: Potassium and naphthalene react to give the naphthalide radical anion, a powerful reducing agent capable of stabilizing reactive intermediates by coordination to the metal prior to CO coordination.

$$VCl_3 + 4\ Na \xrightarrow[\text{diglyme, }-NaCl]{\text{300 bar CO}} [Na(diglyme)_2]^+[V(CO)_6]^-$$
Colourless, diamagnetic

$\downarrow$ H_3PO_4, - H_2

$V(CO)_6$
Green-black, air-sensitive,
paramagnetic (17 VE)

$$CrCl_3 + Al \xrightarrow[C_6H_6,\ AlCl_3]{\text{300 bar CO}} Cr(CO)_6 + AlCl_3$$
Colourless,
diamagnetic, air-stable (18 VE)

$$Mn(OAc)_2 + 10\ CO \xrightarrow[-\ Al(OAc)_3]{AlEt_3,\ Pr^i_2O} Mn_2(CO)_{10}$$
Yellow crystals,
diamagnetic (18 VE)

$$CoCO_3 + 2\ H_2 + 8\ CO \xrightarrow[130°C]{\text{300 bar, CO}} Co_2(CO)_8 + 2\ CO_2 + 2\ H_2O$$
Orange-brown,
diamagnetic (18 VE)

$$Ru(acac)_3 + 4.5\ H_2 + 12\ CO \xrightarrow[130°C]{\text{300 bar CO}} Ru_3(CO)_{12}$$
Yellow, diamagnetic,
air-stable (18 VE)

The structures of metal carbonyls

The structures of metal carbonyls can be explained using the 18-electron rule. This is fulfilled if Cr, Fe and Ni bind six, five and four CO ligands, respectively, to give monomeric compounds. The carbonyls of Mn and Co form metal–metal bonded dimers to attain an 18 VE configuration. The paramagnetic 17 VE complex $V(CO)_6$ is an exception: in order to achieve an electron count of 18, it would have to dimerize to give a 7-coordinate species; such an arrangement is however sterically unfavourable. For similar reasons $Ti(CO)_7$ cannot be prepared, although the isoelectronic complexes $Ti(CO)_5(dmpe)$ (dmpe = $Me_2PCH_2CH_2PMe_2$) and $[Ti(CO)_6]^{2-}$ exist.

Octahedral,
ν_{CO} 2000 cm^{-1}

Octahedral,
ν_{CO} 2044m, 2013s, 1983 cm^{-1}

Trigonal bipyramid,
ν_{CO} 2034s, 2013vs cm^{-1}

D_{3d}, in solution,
ν_{CO} 2107, 2069, 2042, 2031, 2023, 1991 cm^{-1}

C_{2v}, solid state,
ν_{CO} 2112, 2071, 2059, 2044, 2031, 2001, 1886, 1857 cm^{-1}

Tetrahedral,
ν_{CO} 2057 cm^{-1}

Metal–CO binding

Classical ligands such as NH_3 are Lewis bases and form donor bonds to Lewis acids such as Ni^{2+} via their lone electron pairs; they do not, however, react with zerovalent metals to give for example $Ni(NH_3)_4$. On the other hand, CO is a very weak base indeed – its protonated form, the formyl cation CHO^+, is extremely unstable – but does form adducts with zerovalent metals. Why?

Qualitatively, a molecular orbital description of CO shows the existence of a carbon-centred lone pair (HOMO) and of degenerate π^* levels (LUMO's).

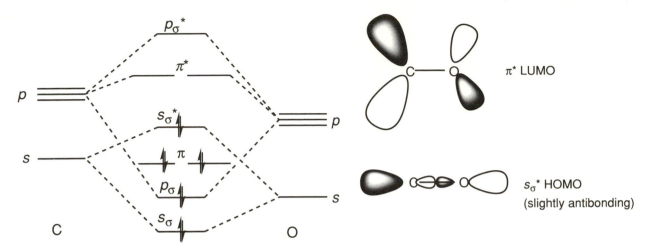

π^* LUMO

s_σ^* HOMO
(slightly antibonding)

While the carbon–lone pair (s_σ^*) interacts with a vacant metal orbital in donor fashion (the most important bonding contribution), the π^* orbital possesses the ideal symmetry for accepting electron density from an occupied metal d-orbital. It is this behaviour as a powerful electron acceptor which stabilizes the M–CO bond in a way that NH_3 is unable to do: *CO is a strong 'π-acid'*. Unsaturated organic ligands in general behave as π-acids, although few have the acceptor strength of CO. This process of electron delocalization over the ligand π^* system is known as **'back-bonding'**.

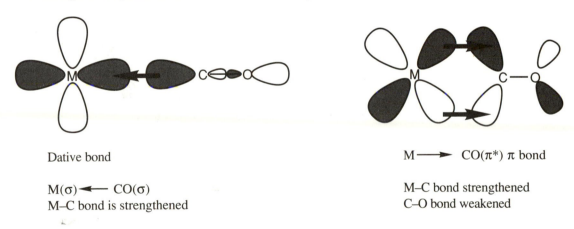

Dative bond

$M(\sigma) \longleftarrow CO(\sigma)$
M–C bond is strengthened

$M \longrightarrow CO(\pi^*)$ π bond

M–C bond strengthened
C–O bond weakened

Another possible bonding contribution, the donation of CO π electrons into a vacant metal d-orbital, is relatively unimportant.

Valence bond resonance hybrids lead to a similar description and indicate a metal–carbon bond order >1 and a C–O bond order <3:

$$\overset{\ominus}{M}-C\equiv O|\overset{\oplus}{} \quad \longleftrightarrow \quad M=C=O\rangle$$

Although the occupancy of the (energetically high-lying) π^* level of CO might be thought to destabilize the system, it has to be remembered that such $2\pi^*$ orbitals are in fact comparable in energy to metal $3d$, $4d$ or $5d$ levels, and therefore back-bonding leads to a substantial overall energy gain. However, if the d-orbitals are contracted (i.e. lowered in energy) by placing positive charges on the metal, as in Ni^{2+}, back-bonding becomes inefficient, and CO is lost. This is illustrated by the disproportionation of $Co_2(CO)_8$ in polar solvents: CO remains bound to Co(–I) only but not to Co^{2+}:

$$Co_2(CO)_8 \quad \xrightarrow{\text{pyridine}} \quad 2\,[Co(py)_4]^{2+}\,[Co(CO)_4^-]_2 \;+\; 8\,CO$$

Bonding modes of CO

In multinuclear complexes CO can adopt doubly and triply bridging coordination modes, recognizable in neutral carbonyl complexes by characteristic ν_{CO} frequencies. A μ_2–CO ligand contributes one electron to the electron count of each metal centre.

Free	Terminal	μ_2	μ_3
ν_{CO} [cm^{-1}] 2143	2120–1850	1850–1750	1730–1620

This is a useful distinction, even though there are numerous borderline cases. It should be noted however that carbonyl anions, where back-bonding is extensive, can have stretching frequencies well below 1800 cm^{-1} although the CO ligands are all terminal.

Asymmetrically bridging CO is also known, as is **isocarbonyl** coordination via the oxygen atom to electron-deficient early transition metals.

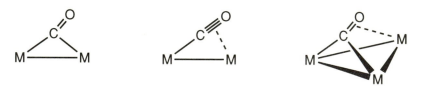

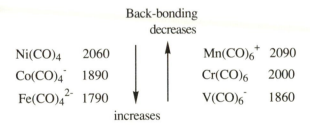

Asymmetric CO bridge

'isocarbonyl' bonding to Ti
(Cp* = C₅Me₅)

IR spectroscopy of metal carbonyls

Since back-bonding depends crucially on the electron density of the metal, it not only strengthens the M–CO bond but provides a very sensitive probe for the electronic characteristics of the metal centre. CO stretching vibrations are accompanied by substantial changes in dipole moment and give rise to intense bands in the infrared spectrum. As we have seen, extensive back-bonding in electron-rich complexes lowers the C–O bond order and thereby the CO stretching frequency, whereas the converse is true if the complex is positively charged or contains other strongly electron-withdrawing ligands. Compare for example:

	ν_{CO} (cm⁻¹)	
CO gas	2143	
$H_3B{\leftarrow}CO$	2164	donor interaction only

In CO the electron pair on carbon is weakly C–O antibonding. If CO binds to a Lewis acid incapable of back-bonding, the influence of the electron pair is effectively removed, and the C–O bond strength (i.e. ν_{CO}) increases.

Back-bonding
decreases

Ni(CO)$_4$	2060
Co(CO)$_4^-$	1890
Fe(CO)$_4^{2-}$	1790

Mn(CO)$_6^+$	2090
Cr(CO)$_6$	2000
V(CO)$_6^-$	1860

increases

As a rule, transition metal carbonyl complexes have CO stretching frequencies of below 2100 cm⁻¹. There are however rare exceptions, such as complexes of electron-withdrawing ligands. In $Pd(C_6F_5)_2(CO)_2$ for example, a square-planar complex with halide-like perfluoroaryl ligands, ν_{CO} is found at 2186 cm⁻¹ – higher than in free CO and an indication that back-bonding in this compound is essentially absent.

Probing the electronic properties of phosphines via ν_{CO}

The sensitivity of the ν_{CO} frequency has been used to quantify the electron donating or withdrawing character of phosphine ligands $L = PR^1R^2R^3$ as a function of the substituents R^1–R^3 (Tolman 1972). Nickel complexes carrying one phosphine ligand, L–$Ni(CO)_3$, are relatively free of steric interactions which might falsify the measurements. The ν_{CO} symmetric (A_1) stretching band is dependent on L:

L in L–$Ni(CO)_3$:	ν_{CO}
PBu^t_3	2056.1
PMe_3	2064.1
PPh_3	2068.9
$P(OMe)_3$	2079.5
$P(OPh)_3$	2085.0
PF_3	2110.8

Each substituent R^1, R^2, R^3 contributes to the frequency shift by a given increment χ, from 0.0 cm^{-1} for $R = Bu^t$ (strongest $+I$ effect, reference standard) to 19.6 cm^{-1} for $R = CF_3$. The frequency shifts induced by introducing one, two, and three substituents on phosphorus are approximately additive:

$$\nu = 2056.1 + \sum_{i=1}^{3} \chi_i$$

This makes it possible to predict the electronic parameters for new phosphine ligands which have not been measured.

In contrast to N and O donors, phosphines are capable of accepting electron density from the metal into the P–C σ^* orbitals. From IR measurements on carbonyl phosphine complexes an order of increasing π-acidity of phosphines is obtained:

$$PBu^t_3 < PMe_3 < P(OMe)_3 < P(OAr)_3 < PCl_3 < CO \approx PF_3$$

Alkyl phosphines are strong donors but poor acceptors, whereas the converse is true for phosphites and phosphine halides. This series may be extended to include non-phosphorus ligands:

$$NH_3 < N\equiv CR < PR_3 < P(OR)_3 < PCl_3 < CO < NO$$

Apart from modifying the electronic characteristics of metal centres, phosphines exert powerful steric influences and thereby control the degree of coordinative saturation and reactivity of a complex. Tolman defined the steric requirement of a phosphine ligand by measuring its cone angle θ:

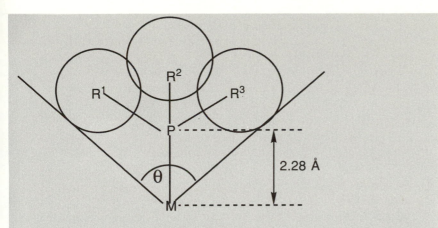

This angle (rather than the C–P–C angles) gives a very vivid illustration of the spacial requirements of phosphine ligands as experienced by the metal centre and is important for our understanding of the reactivity of phosphine complexes, and in particular of homogeneously catalysed processes.

	θ/deg		θ/deg
PH_3	87	PPh_3	145
PF_3	104	$P(OPh)_3$	128
$P(OMe)_3$	107	PPr^i_3	160
PMe_3	118	PBu^t_3	182
PMe_2Ph	122	$P(o\text{-tolyl})_3$	194
$PMePh_2$	136		

Substitution reactions

CO ligands can be substituted by other ligands either thermally or photochemically. Carbonyl substitution of 18 VE complexes follows a dissociative process which generates a coordinatively unsaturated intermediate:

$$L_nM\text{–CO} \underset{+\text{ CO, } k_{-1}}{\overset{-\text{ CO, } k_1}{\rightleftharpoons}} L_nM\text{–}\square \xrightarrow{L', k_2} L_nM\text{–}L'$$

Vacant coordination
site

Second-row complexes usually react faster than either first or third-row homologues–this is one of the reasons why the most active homogeneous catalysts are usually found among second-row elements. Weakly coordinating ligands (ethers, nitriles) facilitate CO removal by stabilizing the coordinatively unsaturated intermediates before they are themselves replaced.

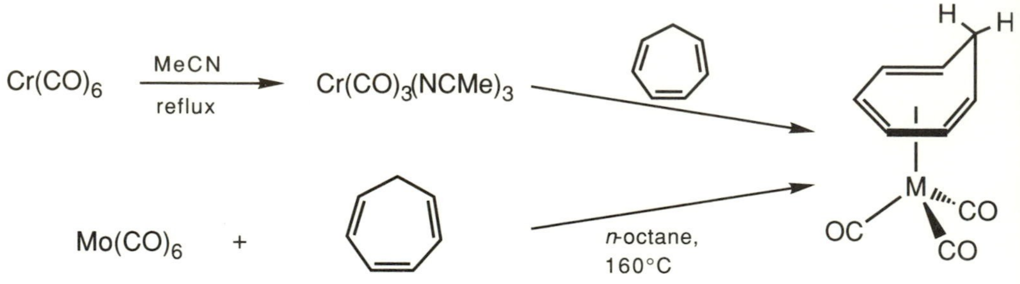

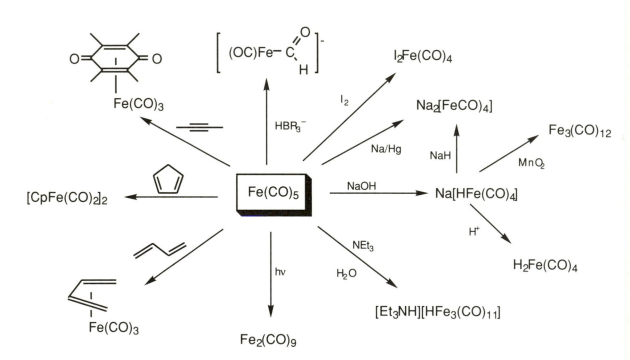

Some of the reactions typical for metal carbonyl complexes are summarized below taking $Fe(CO)_5$ as an example:

Photochemical substitution is assisted by removal of the liberated CO ligand from the reaction equilibrium by purging with nitrogen.

Photochemical substitution usually involves milder reaction conditions and allows the synthesis of less highly substituted or more labile products. Complexes of labile THF, nitrile or alkene ligands are convenient synthons in organometallic chemistry:

$$W(CO)_6 \xrightarrow[\text{THF}]{hv} W(CO)_5(THF) \xrightarrow{L} W(CO)_5(L)$$

$$Fe(CO)_5 \xrightarrow[\textit{cis}\text{-cyclooctene}]{hv} \left\langle \!\! \bigcirc \!\! \right| - Fe(CO)_4 \xrightarrow[\textit{cis}\text{-cyclooctene}]{hv} \left(\left\langle \!\! \bigcirc \!\! \right| \right)_2 Fe(CO)_3$$

Associative substitution can take place with coordinatively unsaturated 4- and 5-coordinate complexes provided that a mechanism exists which avoids an electron count of more than 18. This is possible, for example, if allyl or nitrosyl ligands are present which can change from a 3- to a 1-electron donor mode:

$$(OC)_4 Mn\!\!=\!\!\bar{N}\!\!=\!\!O \xrightarrow{L} (L)(OC)_4 Mn\!-\!N\!\!\diagdown^{O} \xrightarrow{-CO} (L)(OC)_3 Mn\!\!=\!\!\bar{N}\!\!=\!\!O$$

18 VE, linear NO   18 VE, bent NO   18 VE, linear NO
                            Electron transfer catalysis.

Odd-electron complexes (17 or 19 VE) react very much faster than even-electron analogues. Such reactive species need only be present in trace quantities and may be generated by electron transfer catalysis:

$$V(CO)_6 \xrightarrow[\text{rate determining step}]{L} L\!-\!V(CO)_6 \xrightarrow[\text{CO}]{\text{fast}} L\!-\!V(CO)_5$$

17VE   19 VE   17VE product

$V(CO)_6$ reacts 10^{10} times faster than $Cr(CO)_6$!

Electron transfer catalysis may be induced by catalytic quantities of oxidizing agents, such as $AgBF_4$, or electrochemically:

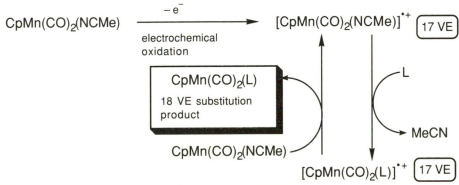

Apparently simple substitution reactions of binuclear complexes with metal–metal bonds involve similar electronically unsaturated radical species through M–M bond homolysis:

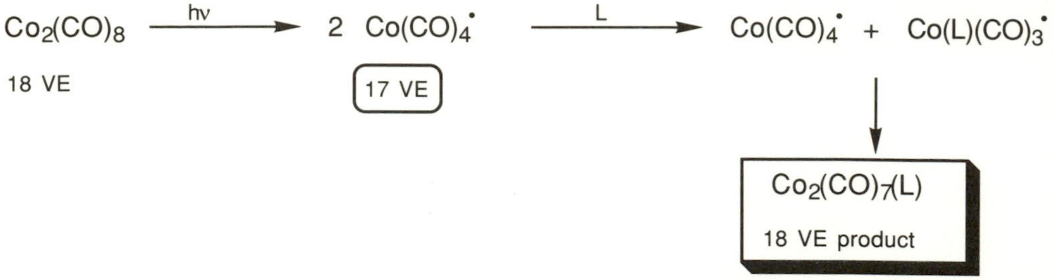

$$Co_2(CO)_8 \xrightarrow{h\nu} 2\ Co(CO)_4^{\bullet} \xrightarrow{L} Co(CO)_4^{\bullet} + Co(L)(CO)_3^{\bullet}$$

18 VE   17 VE

$$Co_2(CO)_7(L)$$

18 VE product

> NOTE: Complete removal of all CO ligands, either by photochemical or thermal methods, is rarely possible. The incoming ligands are usually weaker π-acids than CO, and back-bonding to the remaining CO ligands increases with each successive CO loss, strengthening the M–CO bond.

2.2 Metal carbonyl anions

The reduction of metal carbonyls gives anionic complexes. The high reactivity of these species have led to the development of an enormously rich chemistry; they are susceptible to attack by electrophiles, they can be oxidized or further reduced and give access to hydride, alkyl, and acyl complexes.

Metal carbonyl anions are accessible by a number of routes.

Synthesis

By disproportionation:

$$3\ Mn_2(CO)_{10} \xrightarrow[120\ °C]{pyridine} 2\ [Mn(py)_6]^{2+}\ [Mn(CO)_5^-]_2$$

By nucleophilic attack:

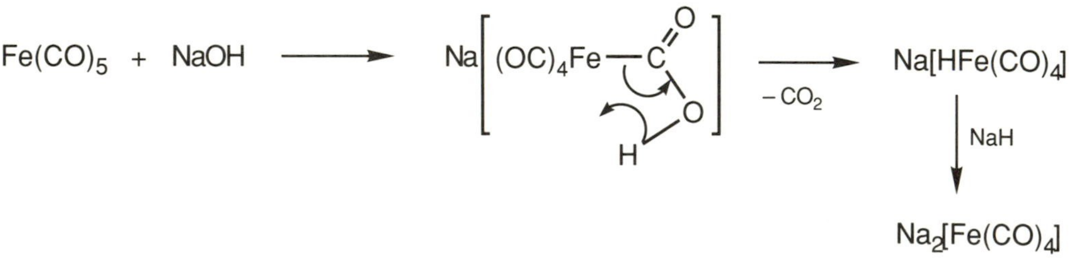

$$Fe(CO)_5 + NaOH \longrightarrow Na\left[(OC)_4Fe-C\begin{matrix}O\\\\O\end{matrix}\ H\right] \xrightarrow{-CO_2} Na[HFe(CO)_4]$$

$$\downarrow NaH$$

$$Na_2[Fe(CO)_4]$$

By reduction:

$$Co_2(CO)_8 + 2\ Na \xrightarrow{\text{THF}} 2\ Na[Co(CO)_4]$$

$$ZrCl_4(THF)_2 + 6\ KC_{10}H_8 \xrightarrow[\text{THF, CO 1 bar}]{\text{15-crown-5}} [K(\text{15-crown-5})]_2\ [Zr(CO)_6]$$

The anions $[M(CO)_6]^{2-}$ (M = Ti, Zr, Hf) are the only binary carbonyls of group IV metals; $Nb(CO)_6^-$ and $Ta(CO)_6^-$ are prepared similarly. The anionic complexes are thermally stable but, in contrast to most neutral metal carbonyls, highly sensitive to oxidation.

Carbonyl metallates are strong bases/nucleophiles. The basicity increases within a group: first < second < third-row metal. $Co(CO)_4^-$ is one of the most stable anions and the least basic; hydrocarbyl ligands greatly increase the basicity.

Relative basicity of carbonyl anions:

$Co(CO)_4^-$	1
$CpCr(CO)_3^-$	4
$CpMo(CO)_3^-$	67
$Mn(CO)_5^-$	77
$CpW(CO)_3^-$	550
$CpNi(CO)^-$	5.5×10^6
$CpFe(CO)_2^-$	7×10^7

While the reduction of metal carbonyls with common reducing agents such as sodium amalgam in THF stops at the stage of mono- or dianions, with stronger reducing agents (sodium in liquid ammonia, lithium–naphthalene, alkali metal and crown ether, or sodium–hexamethylphosphoramide) '**super-reduced' carbonyl metallates** are obtained in which the formal oxidation state of the metal may be as low as –IV, for example: $M(CO)_5^{3-}$ (M = V, Nb, or Ta), $M(CO)_4^{4-}$ (M = Cr, Mo, or W), and $M(CO)_4^{3-}$ (M = Mn or Re):

Metals in oxidation states –III and –IV

$$Na[V(CO)_6] + 3\ Na \xrightarrow{\text{NH}_3\ \text{liqu.}} Na_3[V(CO)_5] + 1/2\ Na_2C_2O_2$$

$$Na[Mn(CO)_5] + 3\ Na \xrightarrow[\text{2. NH}_3\ \text{liqu.}]{\text{1. HMPA}} Na_3[Mn(CO)_4] + 1/2\ Na_2C_2O_2$$

$$K[Co(CO)_4] + 3\ K \xrightarrow{\text{NH}_3\ \text{liqu.}} K_3[Co(CO)_3] + 1/2\ K_2C_2O_2$$

Note that during these reactions CO is reduced to give acetylenediolate, $C_2O_2^{2-}$ as the by-product. Because of very strong back-bonding in these electron-rich complexes, the C–O stretching frequencies are low:

	ν_{CO}/cm^{-1}
$V(CO)_5^{3-}$	1807(w), 1630(s), 1580(s)
$Mn(CO)_4^{3-}$	1790(w), 1600(vs)
$Cr(CO)_4^{4-}$	1657(w), 1462(vs)
$Co(CO)_3^{3-}$	1600(vs)

As may be expected, these are highly reactive and are able to reduce CO_2:

$$M(CO)_5^{2-} + 2\ CO_2 \longrightarrow M(CO)_6 + CO_3^{2-}$$

Metal carbonyl anions in organic synthesis

As nucleophiles carbonyl metallates react readily with organic electrophiles to give alkyl, aryl or acyl complexes. A particularly convenient reagent which is easily accessible on a large scale is $Na_2[Fe(CO)_4]$; its alkyl derivatives have found numerous synthetic applications and it resembles Grignard reagents *(Collman's reagent)*:

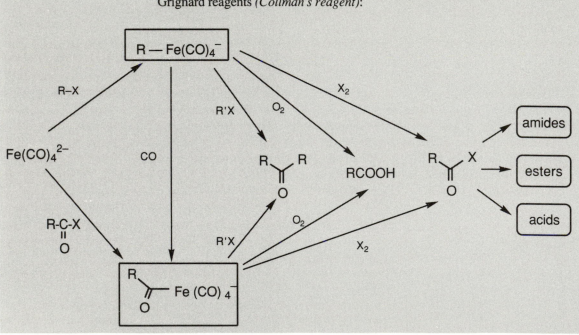

2.3 Metal carbonyl hydrides

Synthesis

By protonation

Protonation of metal carbonyl anions leads to metal carbonyl hydrides. The first transition metal hydride complexes to be isolated were prepared in this way:

$$Co(CO)_4^- + H^+ \longrightarrow HCo(CO)_4$$

$$HFe(CO)_4^- + H^+ \longrightarrow H_2Fe(CO)_4$$

These monomeric carbonyl hydrides are volatile liquids which are stable under an atmosphere of CO and at low temperature. In the absence of CO they decompose to metal carbonyls with liberation of H_2. The thermal stability of carbonyl hydrides increases significantly for the heavier elements within a group.

Neutral carbonyl complexes are protonated by strong acids to give cationic hydrides:

$$Fe(CO)_3(PPh_3)_2 + H_2SO_4 \longrightarrow [HFe(CO)_3(PPh_3)_2]^+[HSO_4]^-$$

Protonation of clusters can lead to 'interstitial hydrogen', i.e. a hydride ligand which is encapsulated by a framework of metal atoms:

$$[Co_6(CO)_{15}]^{2-} + H^+ \longrightarrow [HCo_6(CO)_{15}]^-$$

By reduction

$$FeI_2(CO)_4 \xrightarrow{NaBH_4} H_2Fe(CO)_4$$

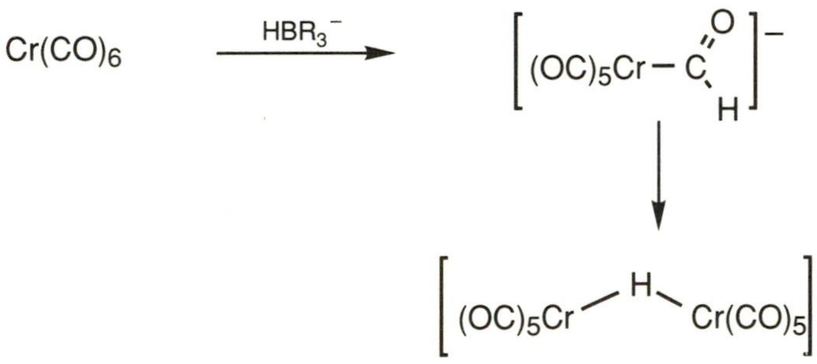

$$Cr(CO)_6 \xrightarrow{HBR_3^-} \left[(OC)_5Cr-\overset{\displaystyle O}{\underset{\displaystyle H}{C}} \right]^-$$

Anionic formyl complex as unstable intermediate

$$\left[(OC)_5Cr \overset{\displaystyle H}{\diagdown} Cr(CO)_5 \right]^-$$

$HCr_2(CO)_{10}^-$ and its Mo and W homologues are rare examples for bent M–H–M bridges unsupported by M–M bonds

From dihydrogen

$$Mn_2(CO)_{10} + H_2 \longrightarrow 2\ HMn(CO)_5$$

$$Co_2(CO)_8 + H_2 \longrightarrow 2\ HCo(CO)_4$$

Transition metal complexes are uniquely able to react with molecular hydrogen by scission of the H–H bond and formation of reactive M–H species, in spite of the high H–H bond enthalpy of 450 kJ mol^{-1}. This 'hydrogen activation' is the basis of several important catalytic reactions, notably hydroformylation and hydrogenation.

The addition of H$_2$ to coordinatively unsaturated metal centres is particularly facile with second– and third–row elements of the Fe, Co and Ni triads and was first illustrated by L. Vaska for iridium:

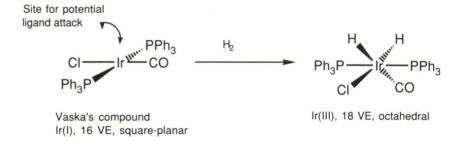

Vaska's compound
Ir(I), 16 VE, square-planar

Ir(III), 18 VE, octahedral

Since the metal changes its formal oxidation state by two units (in this case from +I to +III since H is more electronegative than Ir), hydrogen activation exemplifies an important reaction principle: **oxidative addition**.

Oxidative addition

Oxidative addition can occur when a metal complex behaves as both a Lewis acid and a Lewis base, and reacts with a molecule A–B under bond cleavage and formation of new M–A and M–B bonds:

$$A—B \ + \ \left(\begin{array}{c}\vdots\end{array}\right) M^y L_x \longrightarrow \begin{array}{c} A \\ \diagdown \\ M^{y+2}L_x \\ \diagup \\ B \end{array}$$

This increases the coordination number of the metal centre by two, and since A and B are usually considered to be more electronegative than M, the formal oxidation state of M increases by two units.

Oxidative addition is facile if

☞ $M^y L_x$ is coordinatively unsaturated. Examples: square-planar 16 VE complexes of d^8 and d^{10} metals: Rh^I, Ir^I, Ni^0, Pd^0, Pt^{II}, and Pt^0.

☞ The metal has an energetically accessible oxidation state M^{y+2}, $Ni^0 \rightarrow$ Ni^{II} and $Pt^{II} \rightarrow Pt^{IV}$ are facile but $Ni^{II} \rightarrow Ni^{IV}$ is not.

Besides H_2, many substrates undergo oxidative additions: HCl, Cl_2 and other halogens and interhalogens, RCOOH, $HSiR_3$, alkyl, aryl, vinyl and benzyl halides, acyls RC(O)Cl and O_2. Substrates with double bonds A=B usually add to the metal with retention of an A–B single bond. For example aldehydes, ketones, alkenes and alkynes, particularly with electron-withdrawing substituents, can undergo reactions which amount to an oxidative addition to the metal:

The stability of the higher oxidation state increases with atomic number within a triad. Pd^{IV} may be borderline: a number of Pd^{IV} alkyls are known to be formed by oxidative addition, though the reaction is readily reversed.

$$Pt^0(PR_3)_4 + \quad \underset{O}{\overset{F_3C\diagdown \diagup CF_3}{\|}} \quad \xrightarrow{-2\,PR_3} \quad \underset{R_3P}{\overset{R_3P}{\diagdown}}Pt\underset{O}{\overset{\overset{F_3C}{|}}{\underset{}{\diagdown}}C{-}CF_3}$$

Of course, the readiness of a metal centre to react with potentially oxidative substrates depends on the nature of the metal and the other ligands. While electron-withdrawing ligands such as CO deactivate the metal, strong electron donors such as PMe_3 raise the energy of metal centred non-bonding electron pairs (and so increase the metal basicity) which can lead to dramatic reactivity increases. Similarly, anionic complexes are more reactive than neutral ones.

*One fundamentally important aspect of the oxidative addition is the stereochemistry of the reaction. The new ligands A and B are bonded in **cis** position with respect to each other (sometimes subsequent rearrangements may obscure this). It follows from the principle of microscopic reversibility that the reverse reaction, the **reductive elimination** of A–B from the oxidized complex, can also only proceed if A and B are mutually in **cis** position. In consequence, square–planar trans-dialkyl compounds MR_2L_2 are more stable with respect to reductive elimination than the corresponding cis isomers.*

$$L_nM^y + \quad \underset{B}{\overset{A}{|}} \quad \underset{\text{reductive elimination}}{\overset{\text{oxidative addition}}{\rightleftarrows}} \quad \underset{B}{\overset{A}{\diagdown}}M^{y+2}L_n$$

An interesting recent application of oxidative addition is the activation of saturated hydrocarbons by adding C–H bonds to metals (see p. 72). For functionalized arenes this process has been known for some time and leads to *ortho*-metallation (see. p. 52).

Properties of metal carbonyl hydrides

The strongly electron-withdrawing character of CO leads to a polarization of the M–H bond in $HM(CO)_n$ compounds. As a consequence metal carbonyl hydrides behave as acids in aqueous solution:

Complex	m.p. /°C	v(M–H) /cm^{-1}	δ^1H /ppm	pK_a	acidity comparable to
$HCo(CO)_4$	–26	1934	–10.7	1	HCl
$HCo(CO)_3(PPh_3)$				7	H_2S
$HMn(CO)_5$	–25	1783	–7.5	7.1	H_2S
$H_2Fe(CO)_4$	–70		–11.2	4.4	acetic acid
$HFe(CO)_4^-$				14	H_2O
$CpW(CO)_3H$	69	1854	–7.5	9.0	boric acid

Some general trends:

☞ Replacing CO with strong donor ligands (Cp, phosphines) sharply decreases the acidity of M–H, hence $HCo(PMe_3)_4$ is an extremely strong base.

☞ Thermal stability of the metal hydride varies thus: first row $\ll$ second row $<$ third row.

☞ Acidity of metal hydrides: first row $>$ second row $\geq$ third row. Bridging hydrides are more acidic than complexes with terminal M–H.

☞ In multinuclear hydride complexes hydrogen can be bonded as:

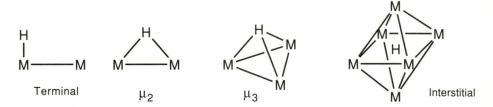

Terminal μ_2 μ_3 Interstitial

Metal hydrides are highly reactive: they insert alkenes and alkynes, are protonated and deprotonated, and acidic hydrides react with diazoalkanes similar to carboxylic acids to give alkyls. They react with chlorocarbons such as CCl_4 to give the corresponding chlorides; this is a useful test reaction for hydrido complexes and is frequently a convenient way of stabilizing otherwise unisolable or unstable intermediates.

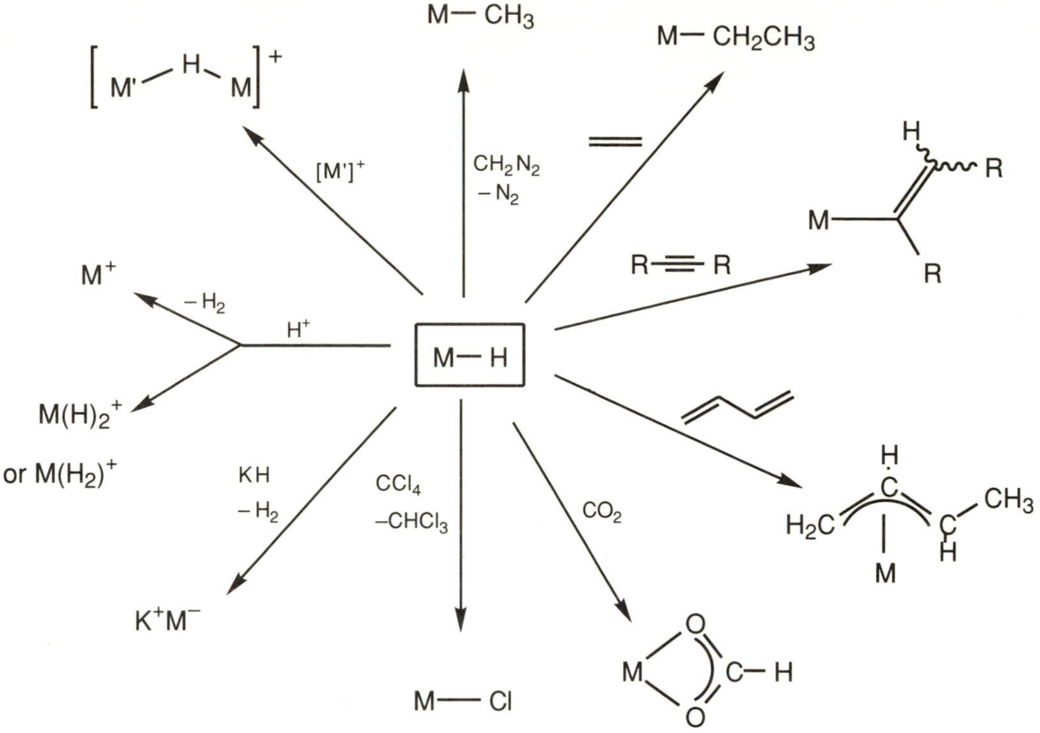

M— CH₃ → $M-CH_3$

$\left[M'\!-\!\overset{H}{}\!-\!M \right]^+$

M— CH₂CH₃ → $M-CH_2CH_3$

$[M']^+$

CH_2N_2
$-N_2$

M^+

$R\!\equiv\!R$

$-H_2$

H^+

$M-H$

$M(H)_2^+$
or $M(H_2)^+$

KH
$-H_2$

CCl_4
$-CHCl_3$

CO_2

K^+M^-

$M-Cl$

One of the most general reactions of transition metal alkyls is the insertion of CO into the M–C bond to give metal acyls. It may appear at first glance surprising that the analogous insertion of CO into M–H to give formyl complexes is not observed:

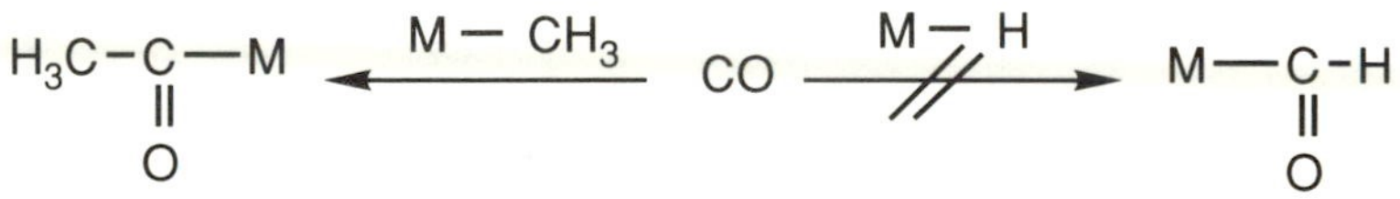

$$H_3C\text{-}\underset{\underset{O}{\|}}{C}\text{—}M \quad\xleftarrow{\; M-CH_3 \;}\quad CO \quad\xrightarrow{\; M-H \;}\!\!\!/\!\!\!/\quad M\text{—}\underset{\underset{O}{\|}}{C}\text{-}H$$

This behaviour reflects the relative stabilities of M–C and M–H bonds: for most metal hydrides, CO insertion would be an endothermic process. For the reaction

$$(OC)_5Mn\text{–}H \;+\; CO \longrightarrow (OC)_5Mn\text{–}CHO$$

$\Delta H \simeq +20$ kJ mol⁻¹. Formyl complexes are, however, accessible by H⁻ attack on coordinated CO (see p. 21). CO insertion into M–H bonds has been discussed as one of the possible mechanisms for the reduction of CO to hydrocarbons (Fischer–Tropsch process, see. p. 33). The possible course of such a reaction and the formation of C–C bonds during the CO reduction process are illustrated by the reaction of Cp*₂ZrH₂ with coordinated CO (Cp* = C₅Me₅).

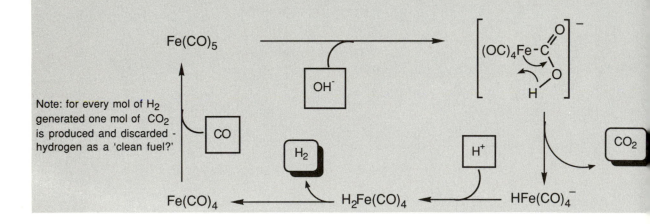

$$Cp^*_2Zr\overset{H}{\underset{H}{\big<}} \quad + \quad Cp^*_2Zr\overset{CO}{\underset{CO}{\big<}} \quad \longrightarrow \quad Cp^*_2Zr-O\underset{H}{\overset{}{\big\backslash}}C=C\underset{H}{\overset{}{\big/}}O-ZrCp^*_2$$

APPLICATION: The water–gas shift (WGS) reaction.

Hydrogen is produced on an industrial scale by the water-gas shift reaction:

$$CO\,(g) \quad + \quad H_2O\,(g) \quad \longrightarrow \quad CO_2\,(g) \quad + \quad H_2\,(g)$$

$$\Delta H^\circ_{298\,K} = -41.2\ kJ\ mol^{-1}$$
$$\Delta G^\circ_{298\,K} = -28.5\ kJ\ mol^{-1}$$

Industrially heterogeneous Fe/Cr or ZnO/Cu catalysts are used; however, metal carbonyls provide useful models and allow a detailed understanding of this reaction. The key step is the nucleophilic attack on a coordinated CO:

Note: for every mol of H_2 generated one mol of CO_2 is produced and discarded - hydrogen as a 'clean fuel?'

$Fe(CO)_5$ OH^- $\left[(OC)_4Fe-C\overset{O}{\underset{O}{\big<}}\right]^-$ H

CO H_2 H^+ CO_2

$Fe(CO)_4 \longleftarrow H_2Fe(CO)_4 \longleftarrow HFe(CO)_4^-$

Dihydrogen complexes

We have seen that dihydrogen adds to low-valent transition metal complexes with cleavage of the H–H bond and formation of two M–H bonds to give a classical hydride complex. This need not always be so. It is possible for H_2 to bind to a metal centre without H–H bond breakage and to form a 'non-classical' H_2 complex. Since the fairly recent isolation of the first example by G. J. Kubas 1984 it has been realized that dihydrogen complexes are not uncommon.

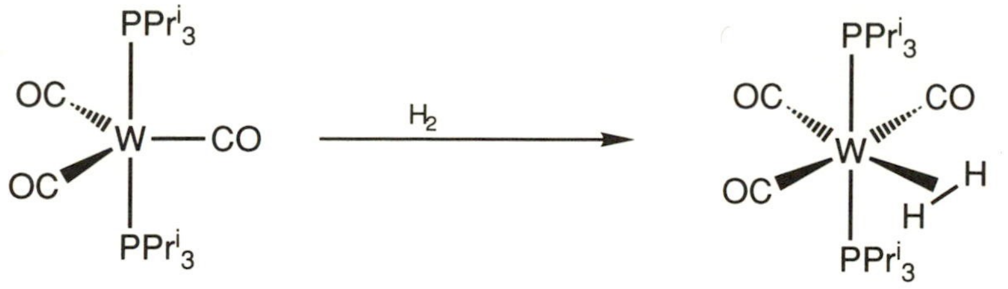

r(H–H) 0.84 Å (neutron diffraction), 0.75 Å (X-ray diffraction); ν_{H-H} 2690 cm^{-1}, $\delta(^1H)$ ca. -4 ppm.

Among the many H_2 complexes now known are those that contain classical M–H bonds as well as M–H_2, e.g. in $[Ir(H)_2(H_2)(PCy_3)_2]^+$. M–H and M–$H_2$ can be distinguished by NMR through measurement of the relaxation time (T_1). Hydrogen in M–H_2 complexes relaxes very much faster than in classical M–H structures. Although this method is not without ambiguities, it appears that minimum times of T_1 <80 ms are indicative of an H_2 complex, while T_1 >150 ms suggest a classical hydride (at 250 MHz).

2.4 Metal carbonyl halides

The introduction of halide ligands is a very facile way of functionalizing metal carbonyl complexes. Metal–metal bonds as are readily cleaved by halogens. The resulting halide complexes are excellent starting materials, e.g. for substitution reactions with nucleophiles.

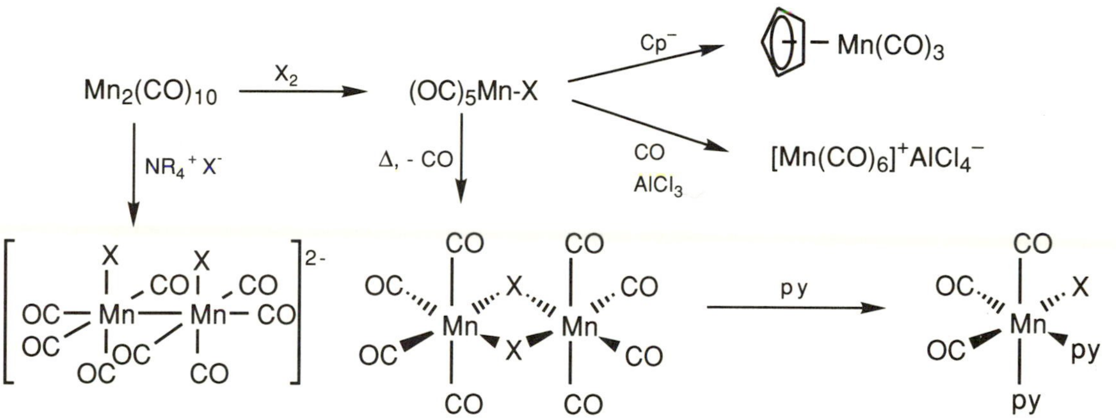

Metal carbonyl halides of noble metals are directly accessible from the metal halides. The alcohol solvent or formaldehyde can act as reducing agents and sometimes serve as the CO source as well:

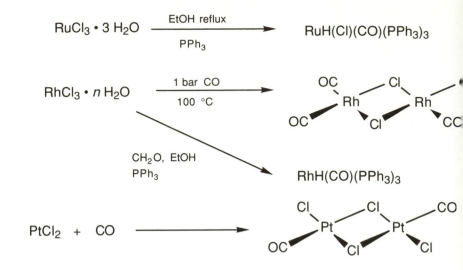

The ruthenium and rhodium complexes in particular are important catalyst precursors. $[PtCl_2(CO)]_2$ was the first metal carbonyl complex ever to be isolated (1870). Cu^I, Ag^I and Au^I, which do not form stable binary carbonyls, also give isolable carbonyl halides.

2.5 Metal carbonyl clusters

The fascinating chemistry of transition metal clusters has grown rapidly over the last 30 years but is beyond the scope of this text. However, some important representatives need to be mentioned here.

Apart from $Fe(CO)_5$, iron forms dinuclear and trinuclear clusters. $Fe_2(CO)_9$ is particularly important; it is involatile, air-stable, and handling is much less hazardous than with $Fe(CO)_5$. $Fe_2(CO)_9$ can be regarded as an adduct of the coordinatively unsaturated fragment $Fe(CO)_4$ with $Fe(CO)_5$; indeed it releases $Fe(CO)_4$ readily and is a valuable starting material for the synthesis of olefin iron carbonyl complexes under non-forcing conditions.

$Fe_3(CO)_{12}$ forms dark green crystals. It contains two bridging CO ligands. In the analogous Ru_3 and Os_3 compounds, all carbonyl ligands are terminal. $Ru_3(CO)_{12}$ and $Os_3(CO)_{12}$ are the simplest carbonyls of the heavier iron homologues which are stable under ambient conditions; $M(CO)_5$ and $M_2(CO)_9$ ($M = Ru$, Os) are only obtained under CO pressure.

☞ Second and third row metals form more stable metal–metal bonds than first row homologues and favour clusters of higher nuclearity.

☞ The larger radii of the heavier elements disfavour CO bridges (but do not rule them out).

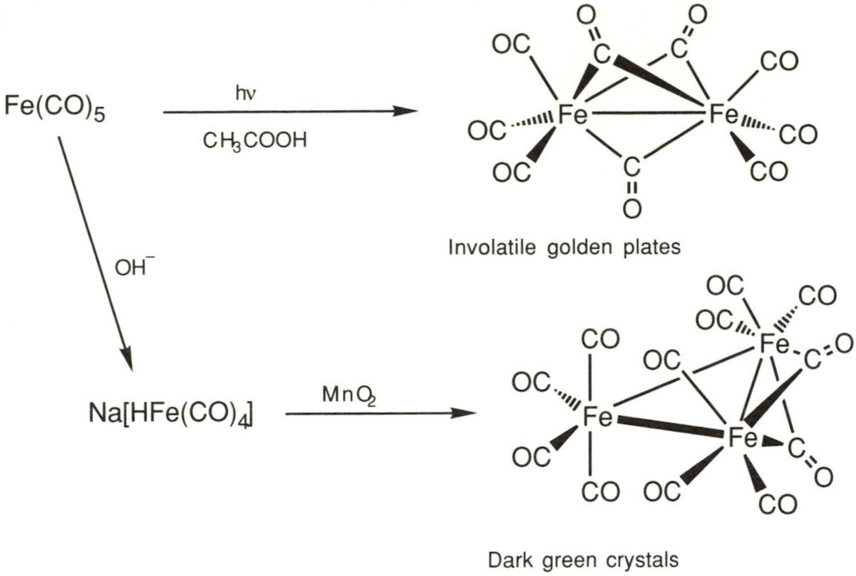

Fe(CO)$_5$ $\xrightarrow[\text{CH}_3\text{COOH}]{h\nu}$

Involatile golden plates

OH$^-$

Na[HFe(CO)$_4$] $\xrightarrow{\text{MnO}_2}$

Dark green crystals

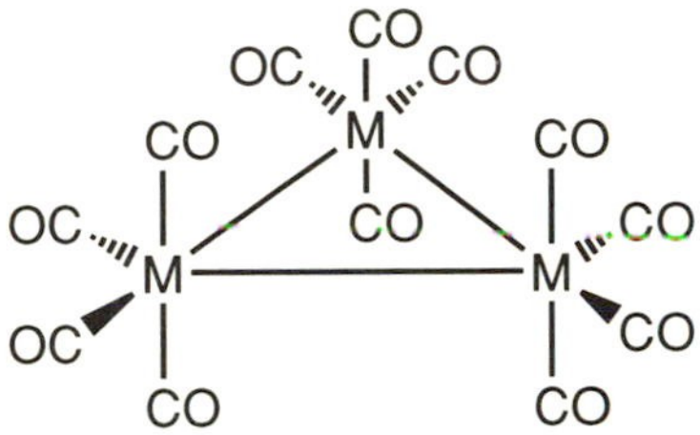

M$_3$(CO)$_{12}$, M = Ru, O

Thermolysis of Co$_2$(CO)$_8$ gives black crystals of Co$_4$(CO)$_{12}$. Following the trends mentioned above, there are Rh$_4$ and Ir$_4$ homologues but no stable M$_2$ compounds. All CO ligands in Ir$_4$(CO)$_{12}$ are terminal.

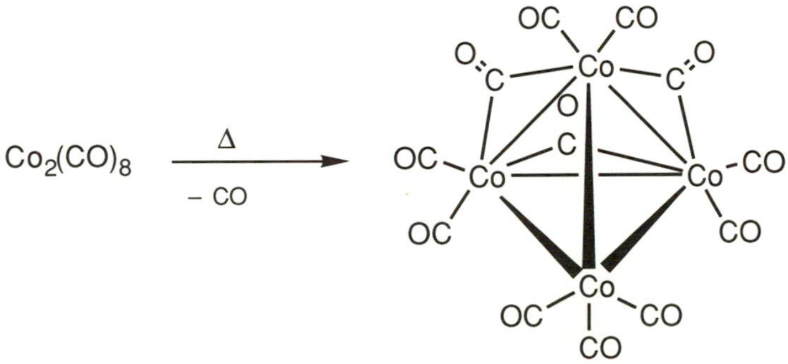

Co$_2$(CO)$_8$ $\xrightarrow[\text{- CO}]{\Delta}$

Cluster chemistry has derived much of its impetus from, among other things, its relationship with metallic surfaces. Larger clusters can frequently be understood as segments of the crystal structure of the pure metal stabilized by CO ligands. The larger clusters are on the borderline between molecular compounds and bulk metal crystallites.

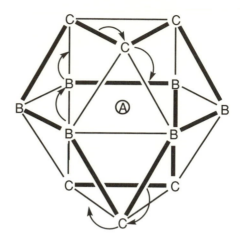

The metal skeleton of $[Rh_{13}(CO)_{24}H_3]^{2-}$ shows the hexagonal close-packing of rhodium atoms. The H atoms migrate freely through the inside of the metal cage, while bridging and terminal CO's exchange over the surface. The arrows show one of three possible pathways for the exchange of bridging and terminal CO's. Bonds with CO bridges are indicated by bold lines.

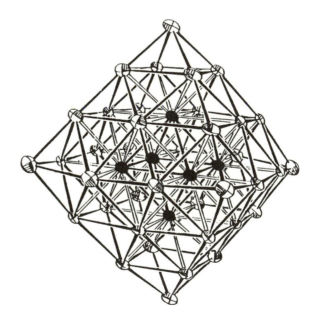

The largest structurally characterized cluster so far: $[Pt_6Ni_{38}(CO)_{48}H_{6-n}]^{n-}$ (n = 4,5). It is a 'cherry cluster', with a 'kernel' of 6 Pt atoms. The CO and H ligands have been omitted for clarity. (Reproduced with permission from G Longoni *et al.* (1985). *Angew. Chem. Int. Ed. Engl.* **24**, 697.)

Wades rules and isolobal relationships

There are several formalisms which rationalize the structure of clusters, and their adequate treatment is beyond the scope of this text. A mention of metal clusters would, however, be incomplete without a brief sketch of Wade's rules.

Wade's rules are a simple way of predicting and rationalizing the structures of metal clusters. The structure of a cluster with n vertices is determined by the number of electron pairs involved in constructing the cluster framework (skeletal electron pairs):

No. of electron pairs	Structure type	Cluster structure
$n + 1$ ➡	*closo*	(n)–vertex polyhedron
$n + 2$ ➡	*nido*	(n)–vertex polyhedron, 1 vertex unoccupied
$n + 3$ ➡	*arachno*	(n)–vertex polyhedron, 2 vertices unoccupied

The basis for this categorization into structural types is the observation that the addition of 2 electrons to a cluster (in the form of 2 e^-, 2 H• or 1 CO) leads to the breaking of one M–M bond. The resulting structure is that of the next largest polyhedron with 1 vertex unoccupied.

Note: It is not sensible to apply Wade's rules to clusters with $n < 4$.

The total number z of cluster electron pairs is given by:

$$z = nd + yl - 12n$$

d = number of d-electrons of each metal atom
n = number of cluster vertices = metal atoms
y = number of ligands
l = number of electrons donated by each ligand

Examples:

Cluster	Number of vertices, n	Cluster electrons	Skeleton electron pairs	Classification
$Os_5(CO)_{16}$	5	$5\times8 + 16\times2 = 72$	$\frac{1}{8}[72 - (5\times12)]$ $= 6$	$n + 1$: *closo*
$Os_5C(CO)_{15}$	5	$5\times8 + 15\times2 + 4$ $= 74$	$\frac{1}{8}[74 - (5\times12)]$ $= 7$	$n + 2$: *nido*

Similarly, the number m of electrons contributed by each vertex metal fragment ML_x carrying x ligands is given by:

$$m = d + xl - 12$$

This leads to a series of isoelectronic cluster fragments. Those with identical numbers m are in principle interchangeable without altering the cluster structure. The isoelectronic relationships between ML_x fragments is a familiar concept:

$Mn(CO)_5^{\cdot}/Re(CO)_5^{\cdot}$; $CpMo(CO)_3^{\cdot}/CpW(CO)_3^{\cdot}$; $Co(CO)_3/Fe(CO)_3^{-}$; $Cr(CO)_3/Cr(C_6H_6)$, etc.

$Mn(CO)_3$:	$7 + (3\times2) - 12 = 1$	FeCp:	$8 + 5 - 12 \qquad = 1$
$Fe(CO)_3$:	$8 + (3\times2) - 12 = 2$	$Ni(CO)_2$:	$10 + (2\times2) - 12 = 2$
$Co(CO)_3$:	$9 + (3\times2) - 12 = 3$	NiCp:	$10 + 5 - 12 \qquad = 3$

On the basis of MO calculations, R. Hoffmann suggested a correlation between the frontier orbitals of metal complex fragments and of hydrocarbyl residues: fragments are isolobal if their frontier orbitals have similar symmetry and shape. An isolobal relationship is symbolized by the sign ⟷.

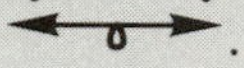

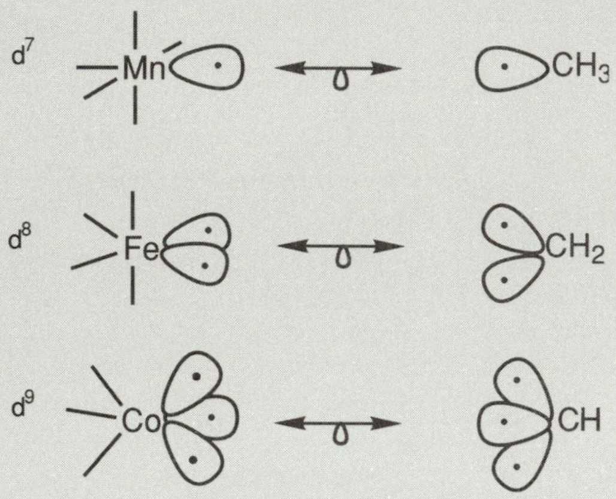

Conversely, ML_x fragments may be isolobal but not necessarily isoelectronic:

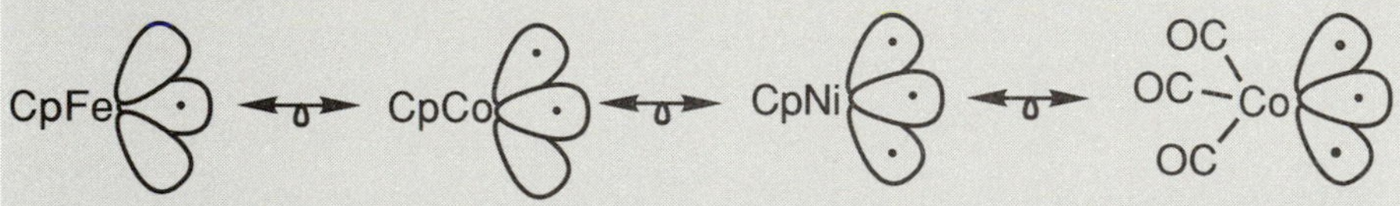

This relationship generalizes our understanding of bonding interactions and is equally valid for organic, organometallic and inorganic systems. Since

it is understandable why compounds as different as cobalt clusters, the hydrocarbon tetrahedrane, and yellow phosphorus have similar structures:

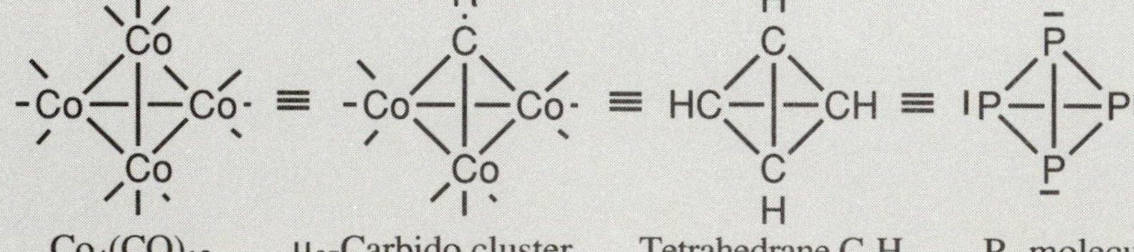

$Co_4(CO)_{12}$ μ_3-Carbido cluster   Tetrahedrane, C_4H_4 P_4 molecule

Similar structural relationships prevail throughout inorganic and organic chemistry, for example $C_5H_5^-$ and the recently isolated P_5^-:

$C_5H_5^-$ P_5^- ferrocene   pentaphosphaferrocene

The Fischer–Tropsch reaction

The Fischer–Tropsch (FT) reaction is the conversion of mixtures of CO and hydrogen ('synthesis gas') into hydrocarbons:

$$n\ CO\ +\ 2n\ H_2\ \longrightarrow\ -(CH_2)_n-\ +\ n\ H_2O$$

The reaction proceeds in the presence of heterogeneous metal catalysts such as iron or ruthenium under high pressure and temperature, and generally gives a mixture of saturated and unsaturated hydrocarbons, as well as products incorporating oxygen such as alcohols and esters ('oxygenates'). The process is attractive, since CO and hydrogen are accessible from any feedstock, such as natural gas, oil, or coal, and supplies a substantial part of the petroleum needs in some countries without natural oil resources, such as South Africa (Sasol process). A related reaction, the hydrogenation of CO to methanol, involves heterogeneous CuO/ZnO catalysts and is carried out on a large scale using for example natural gas (methane) as the CO source. The production of methanol is the most important synthesis gas reaction.

Metal clusters are frequently seen as models for metal surfaces and have played an important role in elucidating many aspects of the complex reaction sequences that are involved in the Fischer–Tropsch process. The conversion of CO into hydrocarbons requires that C–O bond cleavage occurs, followed by C–H and C–C bond formation on the surface of the catalyst, and finally hydrogenolysis and removal of the product into the gas phase:

Metal clusters are able to model several aspects of this reaction sequence. For example, it has long been recognized that thermolysis of metal carbonyls leads to carbido clusters, where the incorporated carbon is derived from a CO ligand. One example is $Fe_5C(CO)_{15}$ where the carbide is located approximately in the base plane of a square pyramid of Fe atoms. Larger clusters with a carbon atom inclosed within the metal skeleton (interstitial carbon) are also known, e.g. $M_6C(CO)_{17}$ (M = Fe, Ru).

$Fe_5C(CO)_{15}$ obeys Wade's rules. For the calculation of the skeletal electron pairs it is useful to consider C as C^{4+}, leaving $Fe_5(CO)_{15}{}^{4-}$ for which the calculation is straightforward.

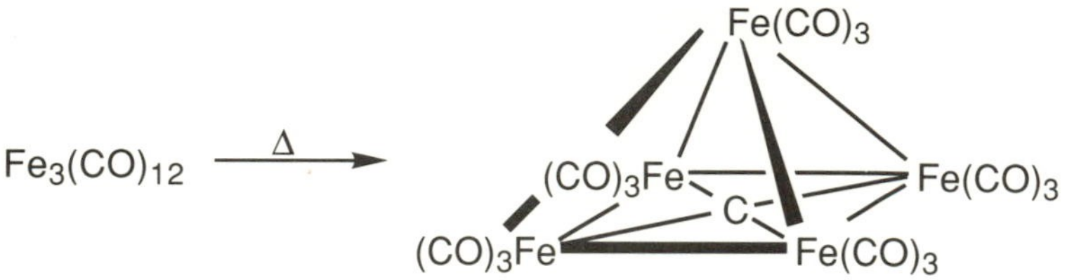

$$Fe_3(CO)_{12} \xrightarrow{\Delta}$$

The reactivity of a surface carbide and its ability to form C–C bonds is exemplified by the following reaction sequence:

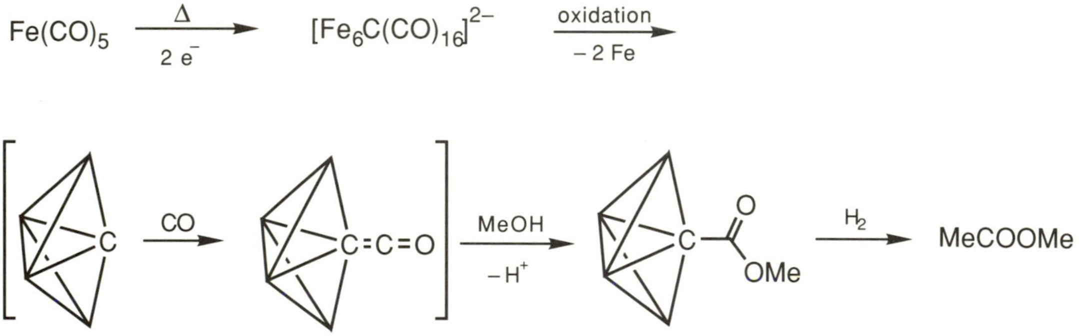

$$Fe(CO)_5 \xrightarrow[2\ e^-]{\Delta} [Fe_6C(CO)_{16}]^{2-} \xrightarrow[-2\ Fe]{\text{oxidation}}$$

$Fe_4(CO)_{12}CCOOMe$

The product, methylacetate, is entirely derived from CO and hydrogen.

The C–C bond forming processes that occur on the surface of Fischer–Tropsch catalysts are illustrated by the coupling of carbene and methyl ligands during the thermolysis of molecular complexes, to give methane and propene as typical Fischer–Tropsch products:

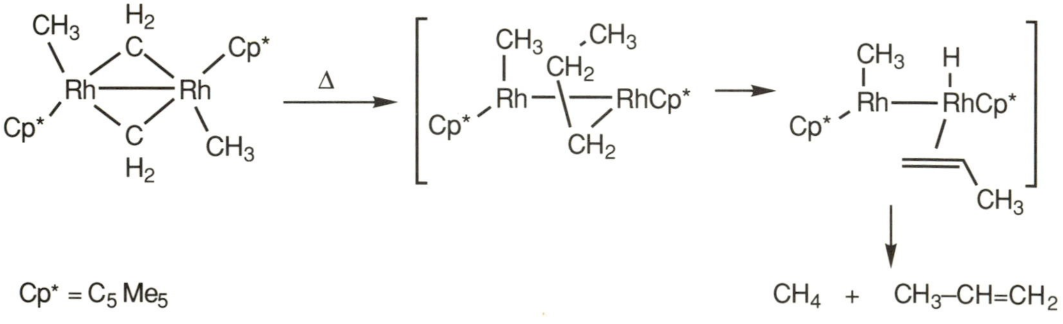

Cp* = C_5Me_5

$$CH_4 \ + \ CH_3–CH=CH_2$$

Another feature of the Fischer–Tropsch reaction, the involvement and interconversion of surface CH_n species, is also reflected in metal cluster chemistry. Note the ease with which metal assemblies are able to make and break C–H bonds:

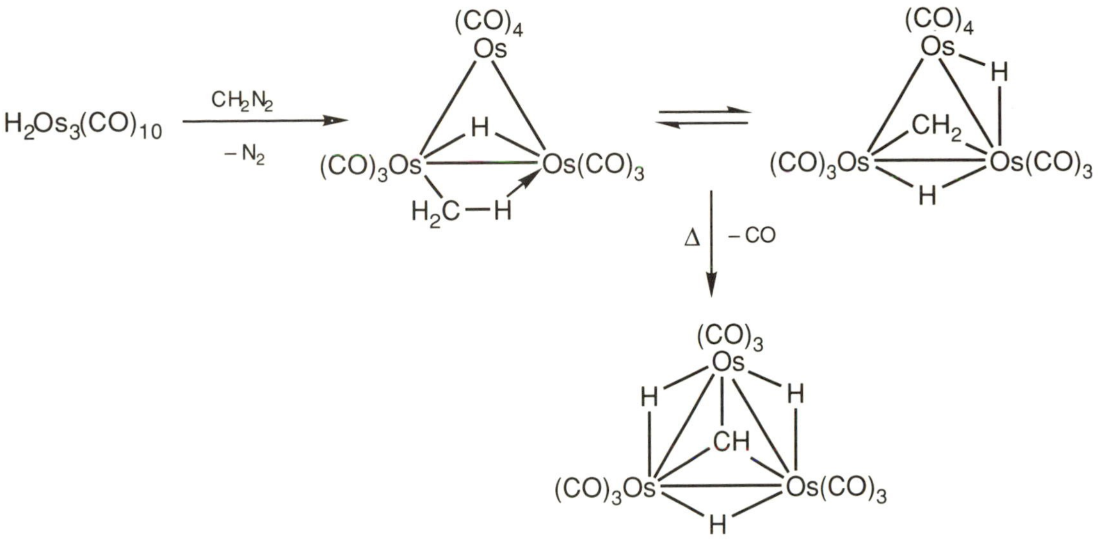

2.6 CS, CSe and CTe complexes

The higher homologues of carbon monoxide: CS, CSe and CTe, are better σ-donors as well as better π-acceptors than CO. The electron pair centred on carbon is significantly C–E antibonding (E = S, Se Te), whereas in CO it has only weak antibonding character. Since CS and CSe, not to mention CTe, are not stable as free molecules, these ligands have to be generated within the coordination sphere of the metal. A common route is via desulfurization of CS_2 with phosphines as sulfur scavengers:

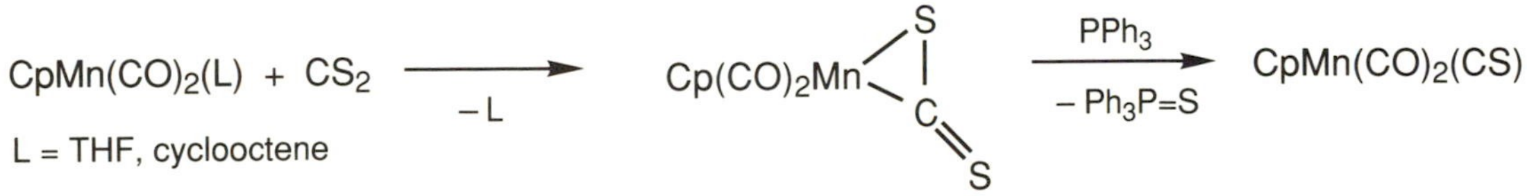

CSe complexes are prepared similarly. Repetition of this procedure allows the isolation of bis- and tris-CS complexes, $CpMn(CO)(CS)_2$ and $CpMn(CS)_3$.

Nucleophilic substitution reactions on coordinated dichlorocarbene ligands have given access to the whole series of CE complexes:

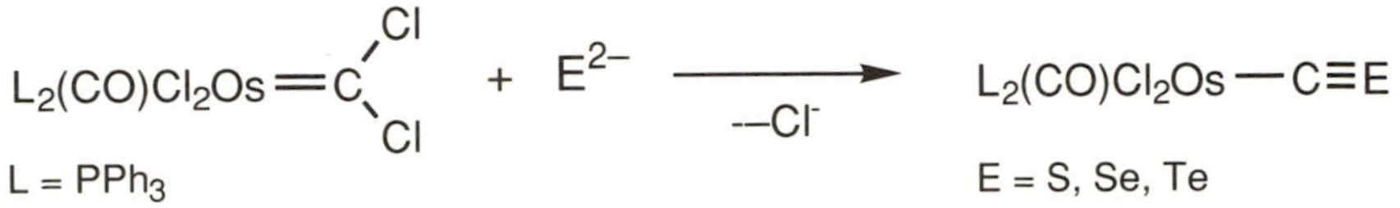

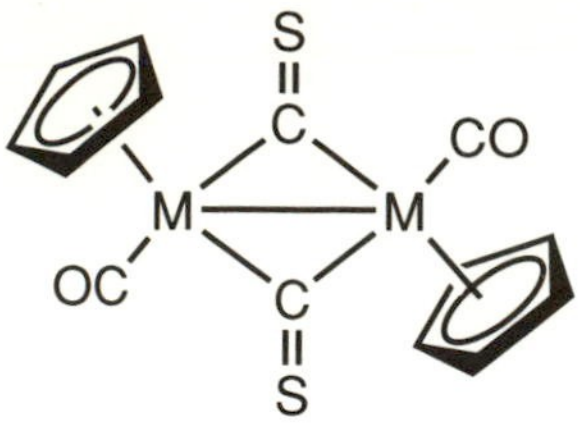

In binuclear mixed CO/CS complexes it is CS which prefers the bridging position since this allows a favourable reduction of the C–S bond order, for example in $[CpM(CO)(\mu–CS)]_2$ (M = Fe, Ru).

Infrared spectroscopy has proved to be an almost ideal tool for studies of metal carbonyl complexes since the C–O stretching frequencies occur in an otherwise clear region of the spectrum. For CS and CSe, IR is less easily applied since their vibrations fall within the fingerprint region of the spectrum and are more difficult to identify.

	$\nu_{C\equiv E}$ /cm^{-1}
M–CS terminal	1410–1160
$M_2(\mu$-CS)	1150–1020
M–CSe terminal	1130–1060

CS ligands are comparatively reactive, with ready cleavage of the C–S bonds, for example:

$$(CO)_5W(CS) + H_2NMe \longrightarrow (CO)_5W–C\equiv NMe + H_2S$$

2.7 Isocyanide complexes

Isocyanides $R–N\equiv C$ are isoelectronic with CO but are stronger σ-donors and weaker π-acceptors; they are therefore able to form stable complexes e.g. with d^0 metals and metals in higher oxidation states where back-bonding is unimportant in cases where there are no CO analogues, e.g. $[Mn(CNR)_6]^{2+}$. Some of these are very stable, even in aqueous media; radioactive technetium isocyanide complexes $[Tc(CNR)_6]^{2+}$ are used in nuclear medicine e.g. for the radio–imaging of the heart.

Isocyanide ligands may be terminal or bridging, bent or linear:

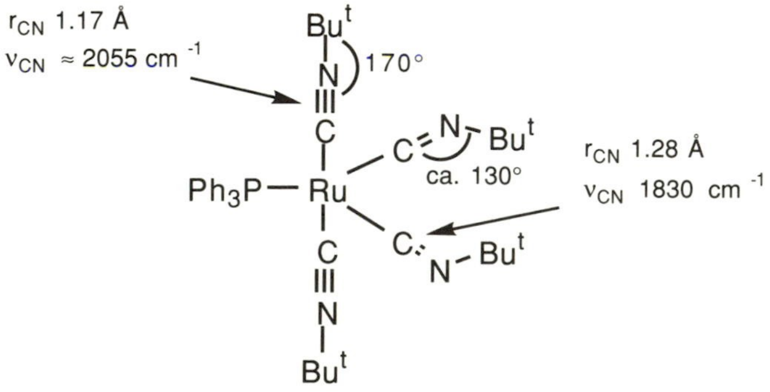

Synthesis

$$Ni(CO)_4 + 4 \text{ PhNC} \longrightarrow Ni(C\equiv NPh)_4 + 4 \text{ CO}$$

$$FeCl_2(CNR)_4 + RNC \xrightarrow{\text{Na/Hg}} Fe(C\equiv NR)_5$$

$$Fe(CNR)_5 \quad \xrightarrow{\text{h}\nu} \quad (RNC)_3Fe(\mu_2\text{-CNR})_3Fe(CNR)_3$$
$$(R = Et, Pr^i)$$

Isocyanides are very reactive and, in contrast to CO, frequently undergo multiple insertions. They are susceptible to attack by nucleophiles to give carbene complexes:

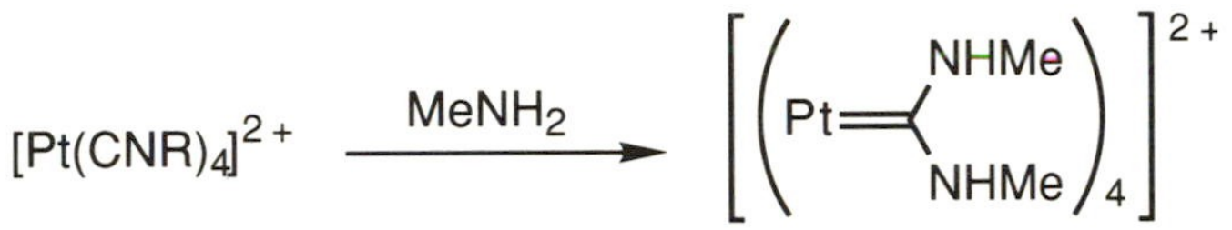

Ni(CNR)$_4$ + MeI $\longrightarrow$

$$[Pt(CNR)_4]^{2+} \xrightarrow{\text{MeNH}_2} \left[\left(Pt=\begin{smallmatrix}NHMe\\ \\NHMe\end{smallmatrix}\right)_4\right]^{2+}$$

2.8 Nitrosyl complexes

Nitric oxide, NO, contains one electron more than CO and can act as a 1– or a 3–electron ligand, depending on its coordination geometry:

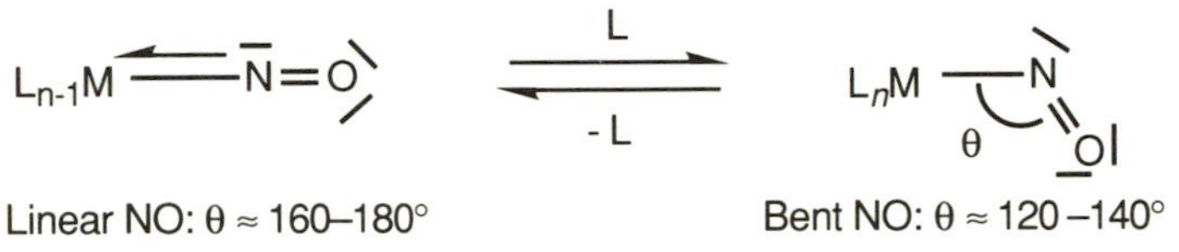

Linear NO: $\theta \approx 160\text{–}180°$ Bent NO: $\theta \approx 120\text{–}140°$

The close relationship between NO and CO is illustrated by the existence of a continuous series of nitrosyl and nitrosyl–carbonyl complexes that are isoelectronic and isostructural with Ni(CO)$_4$. Two NO ligands can replace three COs.

Cr(NO)$_4$ Mn(NO)$_3$(CO) Fe(NO)$_2$(CO)$_2$ Co(NO)(CO)$_3$ Ni(CO)$_4$

The unpaired electron of free NO resides in a π^* orbital and is easily removed to give NO$^+$ (as in NO$^+$PF$_6^-$) which is isoelectronic to CO but a stronger π-acceptor. The antibonding character of the unpaired electron is reflected in the IR stretching frequencies:

	v_{NO} /cm^{-1}
Free NO	1876
Free NO$^+$	2250
M–NO	1900–*ca.* 1600

The v_{NO} frequencies in nitrosyl complexes vary widely and cannot be used as reliable indicators for a bent or linear coordination mode of the nitrosyl ligand.

NO complexes are usually made from NO or NO$^+$ salts and have been known for a long time. The nitroprusside ion $[Fe(CN)_5(NO)]^{2-}$ is used as an analytical reagent in qualitative spot tests for reducing substrates with which it gives characteristically coloured products, such as HS$^-$, S^{2-} or SO$_3^{2-}$.

Metal clusters with doubly or triply bridging NO are also known, for example:

Oxidation of $[Fe(CN)_6]^{3-}$ gives the nitroprusside anion:

$$[Fe(CN)_6]^{3-} \xrightarrow{HNO_3}$$

$$[Fe(CN)_5(NO)]^{2-}$$

Coordinated NO may by reduced to an amido ligand:

$$CpCr(NO)_2Cl \xrightarrow{BH_4^-} Cp_2Cr_2(NO)_4 \; +$$

2.9 Dinitrogen complexes

Although not strictly an organometallic ligand, N$_2$ is isoelectronic to CO, and provides an interesting comparison. It lacks the polarity of CO and is both a very much weaker donor and a weaker π-acceptor. As a consequence, many N$_2$ complexes are labile, and the N–N bond distance of coordinated N$_2$ differs in most cases remarkably little from that of free N$_2$ (1.0976 Å).

N$_2$ coordinates usually end-on as a terminal ligand. Side-on (η^2) coordination and bridging N$_2$ are rare:

M⚌N≡N   M⚌N≡N⚌M

$$\begin{array}{c} N \\ M \Vert \\ N \end{array}$$ $$\begin{array}{c} N \\ M M \\ N \end{array}$$

Binary complexes between metals and N_2 are unstable; e.g. $Ni(N_2)_4$ has been identified by matrix isolation techniques and decomposes at low temperatures. The first dinitrogen complex was isolated in 1965, $[Ru(NH_3)_5(N_2)]^{2+}$, made from $RuCl_3$ and hydrazine. In most dinitrogen complexes known to date, the metal is in a low oxidation state, and back-bonding, though weak, is important for the stability of the complexes.

A number of bacteria are able to use N_2 and reduce it to ammonia, the best known example being *Rhizobium* which is found in the root nodules of legumes and is responsible for the biological fixation of about 100–200 million tons of nitrogen annually worldwide (approximately twice the amount of ammonia produced industrially by the Haber–Bosch process). This is remarkable in view of the very high dissociation energy of N_2 (945.4 kJ mol^{-1}). The bacterial enzyme responsible, nitrogenase, contains redox-active Fe_4S_4 cubane clusters as well as molybdenum – the only case where nature has made use of a second row metal. Consequently there has been intense interest over the years in dinitrogen complexes, particularly of Group 6 and related metals, in order to elucidate the mechanism of nitrogen fixation.

Dinitrogen complexes are most readily prepared by reduction under a nitrogen atmosphere:

$$MCl_4L_2 + 2\,N_2 \xrightarrow[+\,2\,L,\,-\,4\,NaCl]{Na/Hg} cis\text{-}M(N_2)_2L_4$$

$$M = Mo, W$$
$$L = PMe_2Ph$$

These complexes decompose on protonation with partial reduction of N_2 to ammonia:

$$cis\text{-}M(N_2)_2L_4 \xrightarrow[\text{2. base distillation}]{\text{1. }H_2SO_4,\,MeOH} N_2 + 2\,NH_3 + \cdots$$

With chelating ligands, such as $Ph_2PCH_2CH_2PPh_2$ (dppe) *trans* complexes are produced:

1.118(8) Å

$$\underset{P \quad\quad P}{\overset{P \quad\quad P}{N\equiv N - Mo - N\equiv N}}$$ 2.041(5)

By contrast, some bridging N_2 complexes of metals which prefer to be in their highest obtainable oxidation state, such as tantalum, are best formulated as complexes of the N_2^{4-} tetraanion, with a significantly elongated N–N bond and short metal–nitrogen distances indicative of double bonds:

$$1.28 \text{ Å} \quad 1.80 \text{ Å}$$
$$(R_3P)(THF)Cl_3Ta \!=\! N \!-\!\!-\!\! N \!=\! TaCl_3(THF)(PR_3)$$

The terminal nitrogen of the N_2 ligand in low–valent complexes such as $W(N_2)_2L_4$ carries a partial negative charge. This is best illustrated by the resonance hybrid formalism and explains the reactivity of these complexes towards electrophiles:

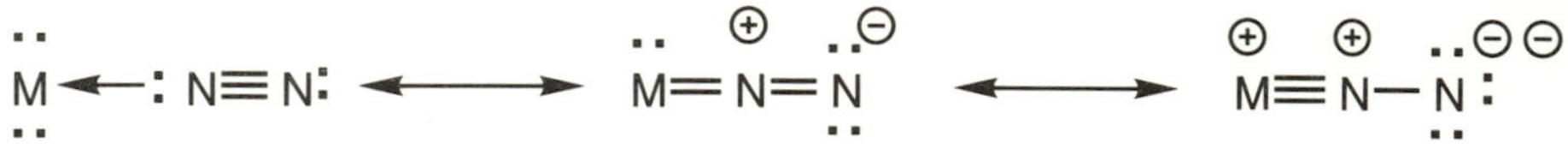

The terminal nitrogen can be functionalized and even incorporated into organic moieties; with conversion of the original N_2 into a hydrazido ligand, $R_2N–N^{2-}$.

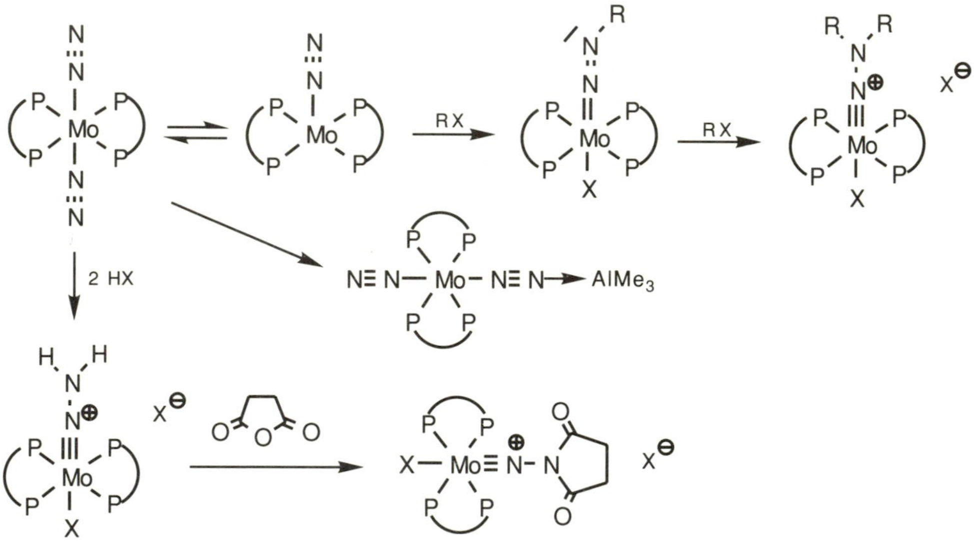

Further reading

H. Werner (1990). Complexes of CO and its Relatives: An Organometallic Family Celebrates its Birthday, *Angew. Chem. Int Ed. Engl.,* **29**, 1077.

C. A. Tolman (1977). Steric Effects of Phosphorus Ligands in Organometallic Chemistry and Homogeneous Catalysis. *Chem. Rev.* **77**, 313.

J. E. Ellis (1990). Highly Reduced Metal Carbonyl Anions, *Adv. Organomet. Chem.* **31**, 1.

G. Wilkinson, F. G. A. Stone, and E.W. Abel (eds.) (1982). *Comprehensive Organometallic Chemistry.* Pergamon, Oxford. A comprehensive treatise in 9 volumes.

3 Metal alkyl complexes

3.1 Stability and structure

The making and breaking of metal–carbon σ-bonds plays an important role in organometallic chemistry and is central to its application to catalysis. Whenever alkanes, alkenes or alkynes are generated, hydrogenated, polymerized or functionalized, metal alkyl intermediates are involved. It is estimated that about three quarters of all products of the chemical industry pass through a catalytic process at some stage.

The first metal alkyls were prepared even before the metal carbonyls, in 1848 when Edward Frankland attempted to stabilize the 'ethyl radical' by reacting zinc with iodoethane and isolated EtZnI and ZnEt$_2$ – a very significant experimental feat at the time in view of the pyrophoric nature and extreme hydrolysis sensitivity of zinc alkyls. Who nowadays would dream of using hydrogen as a protective atmosphere?

The synthesis of zinc alkyls was soon followed by the isolation of alkyl compounds of many main group elements, such as aluminium, magnesium, mercury, silicon, tin and lead. The first transition metal alkyls were not prepared until 1907 when W. J. Pope reported PtMe$_3$I whose tetrameric, cubic structure with octahedrally coordinated Pt^{IV} was not recognized until some 50 years later. For many years, most attempts to prepare other transition metal alkyls similar to the compounds of main group elements failed, and as late as the 1950s it was argued (and indeed supported by theoretical considerations) that transition metal-to-carbon σ-bonds were intrinsically unstable, particularly if 'stabilizing' π-ligands were absent.

This situation changed in the 1960s. A large number of metal–alkyl complexes, with and without π-ligands, are now known, and in many cases the M–C bond enthalpies have been determined. Below are the bond dissociation enthalpies D for a series of main group and transition metal compounds:

It is important to be clear what is meant by 'stability'. Many compounds can be heated to high temperatures without decomposition under inert gas but react violently on exposure to air or moisture. Others, whose thermodynamic data indicate lability, nevertheless appear perfectly stable to oxygen or water. The term 'stability' is used here to indicate the lack of decomposition of the pure compound, i.e. in the absence of a reaction partner.

Main group alkyl	ΔH_f/kJ mol^{-1}	D/kJ mol^{-1}	Transition metal alkyl	D/ kJ mol^{-1}
CMe$_4$	−167	358	Ti(CH$_2$Bu^t)$_4$	198
SiMe$_4$	−245	311	Zr(CH$_2$Bu^t)$_4$	249
GeMe$_4$	−71	249	Hf(CH$_2$Bu^t)$_4$	266
SnMe$_4$	−19	217	TaMe$_5$	261±5
PbMe$_4$	+136	152	WMe$_6$	160±6

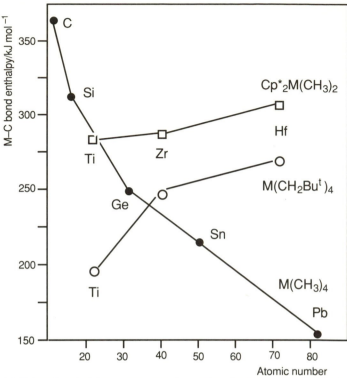

Mean M–C bond enthalpies versus atomic number for representative main group and transition elements (Cp* = C_5Me_5).

General trends:

☞ The M–C bond enthalpy (bond strength) *decreases* with increasing atomic number for main group elements but *increases* within a transition metal triad.

☞ The M–C bond enthalpy of transition metal alkyls (150–300 kJ mol^{-1}) is equal to that of main group elements and comparable to the strength of the carbon–iodine bond.

On thermodynamic grounds, therefore, transition metal alkyls should be much more stable than, for example, lead alkyls which are endothermic and possess only weak Pb–C bonds. This is in sharp contrast to the observed decomposition temperatures of comparable transition metal and lead alkyls:

	Decomposition temperature		Decomposition temperature
$TiMe_4$	> −50°C	$PbMe_4$	> 200°C (b.p. 110°C)
$TiEt_4$	existence doubtful	$PbEt_4$	> 100°C

Bond dissociation enthalpies of transition metal complexes stabilized by donor ligands.

Complex	D/kJ mol^{-1}	Complex	D/kJ mol^{-1}
Cp*$_2$TiMe$_2$	281±8	(CO)$_5$Mn–CH$_3$	187±4
Cp*$_2$ZrMe$_2$	284±2	(CO)$_5$Mn–Ph	207±11
Cp*$_2$HfMe$_2$	306±7	(CO)$_5$Mn–C(O)CH$_3$	160±10
Cp$_2$MoMe$_2$	166±8	(CO)$_5$Re–CH$_3$	220±11
Cp$_2$WMe$_2$	221±3	(CO)$_5$Mn–Cl	294±10
cis-(Et$_3$P)$_2$Pt(Cl)Me	251±30	(CO)$_5$Mn–Br	242±6
trans-(Et$_3$P)$_2$Pt(Cl)Et	206	(CO)$_5$Mn–I	195±6
		MeI	238±1
(Cp = C$_5$H$_5$; Cp* = C$_5$Me$_5$)		MeC(O)I	209±3

The reason for the difference in stability between the tetraalkyls of titanium and lead is therefore not thermodynamic but *kinetic*. In the case of PbEt$_4$ all available orbitals are occupied: the four σ-bonds to carbon supply an electron octet, the $5d$ shell is filled, the $6d$ and σ* levels are energetically inaccessible. As a result, PbEt$_4$ can only decompose by M–C bond homolysis, a comparatively high energy process:

Kinetic versus thermodynamic stability.

$$PbEt_4 \xrightarrow{\Delta} PbEt_3^{\bullet} + Et^{\bullet}$$

By contrast, transition metals such as Ti possess vacant $3d$ orbitals which provide a facile decomposition pathway by allowing interaction with C–H bonds of the alkyl ligands. Since a hydrogen from a carbon in β-position to the metal is abstracted, this process is termed β-**hydrogen elimination** (or, simply, β-**elimination**):

The resulting Ti(H)(Et) species is itself highly unstable:

Determination of the alkene/alkane ratio by GC analysis of the gas phase gives an indication of the decomposition mechanism.
The end product is usually finely divided metal or $M(C_xH_y)$.

This release of alkane amounts to a reduction of the metal by two units: this is a **reductive elimination** step. The process is concerted and can only proceed if the hydride and alkyl ligands are mutually *cis*. The overall β-elimination reaction sequence results therefore in the formation of an alkene and an alkane in equimolar amounts.

Reductive elimination *without* β-H abstraction is a favoured decomposition pathway for some metal dialkyls, e.g. of palladium. Saturated hydrocarbons with double the number of carbons of the alkyl ligands are produced, but no alkenes.

Reductive elimination

☞ The two ligands to be coupled have to be mutually *cis*.
☞ the elimination process is *concerted* and does not involve M–C bond homolysis (cf. $PbEt_4$).

The synthesis of stable transition metal alkyls can be achieved by applying two strategies:

1. *Blocking any available coordination sites by donor ligands.*

This method is widely applied and successful, for example using chelating ligands, such as 2,2'-bipyridyl (bipy) or the strongly basic diphosphine $Me_2PCH_2CH_2PMe_2$ (dmpe). Whereas $TiMe_4$ decomposes above –50 °C, $TiMe_4(dmpe)$ is stable at room temperature. There are also numerous alkyl complexes with CO and Cp ligands which obey the 18-electron rule and are very stable, e.g. $CpFe(CO)_2R$ (R = Me, Et, etc.).

2. *Using alkyl ligands stable to β-hydrogen elimination.*

Frequently used examples of alkyls without β-hydrogens are:

Similarly, stable complexes are obtained from alkyl ligands which possess β-hydrogens but where β-elimination is energetically unfavourable, such as bridgehead alkyls:

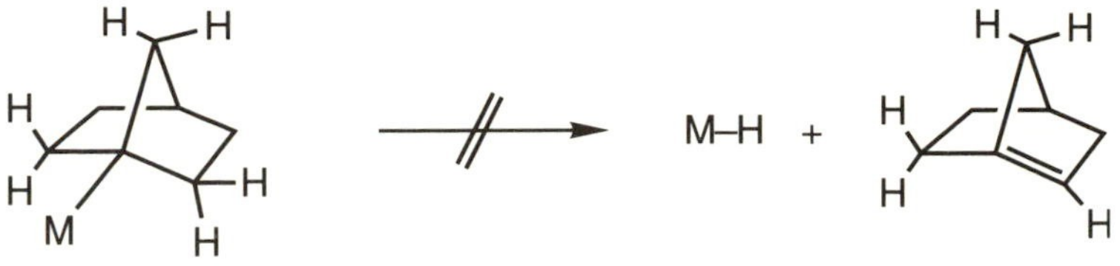

Bredt's rule: Double bonds to bridgehead carbons are unfavourable

The same generally applies to aryl ligands (although a number of aryne complexes are now known):

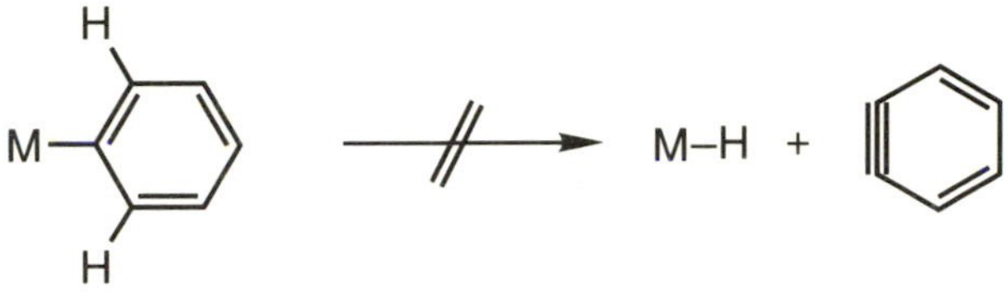

Arynes cannot be formed from aryl ligands with *ortho*-substituents, such as mesityl (2,4,6-Me$_3$C$_6$H$_2$–). Pentafluoro- and pentachlorophenyl ligands belong to this category; they are strongly electron-withdrawing and behave much like halide anions.

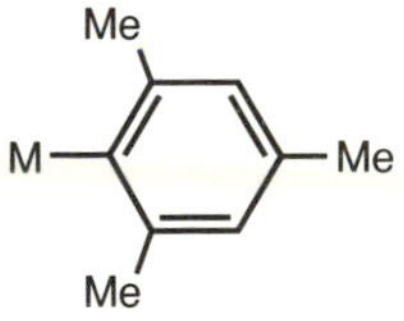

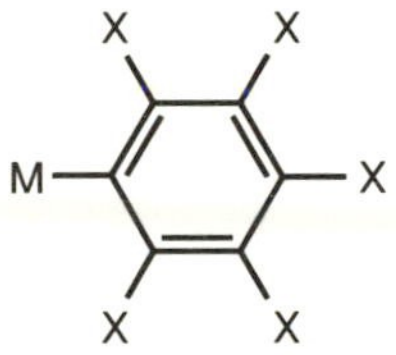

X = F, Cl

The metal–carbon bond strength increases in general with increasing *s*-character of the M–C bond:

$$M-CR_3 \quad < \quad M\text{–aryl}, \quad M\text{–vinyl} \quad < \quad M\text{-}C\equiv C\text{-}R$$

$$sp^3\text{–C} \quad < \quad sp^2\text{–C} \quad < \quad sp\text{–C}$$

bond strength / electron-withdrawing character of C

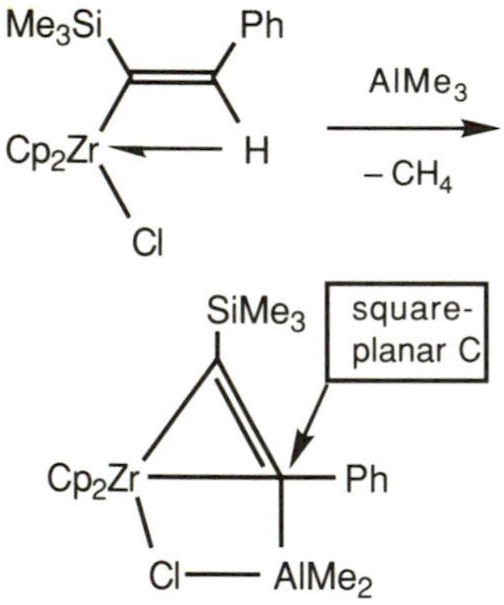

The following metal–alkyl bonding modes are commonly encountered:

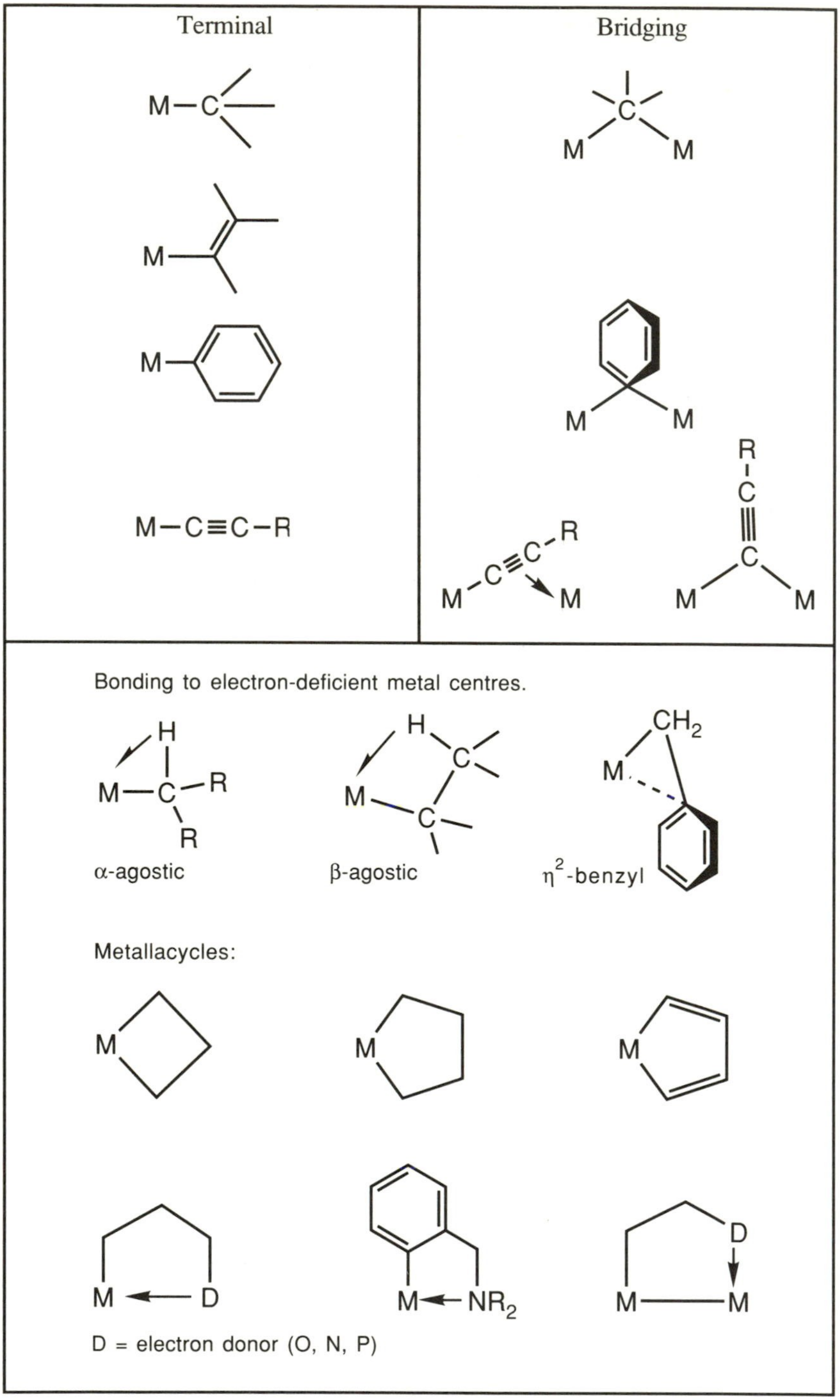

Methyl ligands can adopt a surprising number of coordination modes, depending on the electronic requirements of the metal centres to which they are bound, including η^3–CH_3 and trigonal–planar C. For C–H interactions see p. 72.

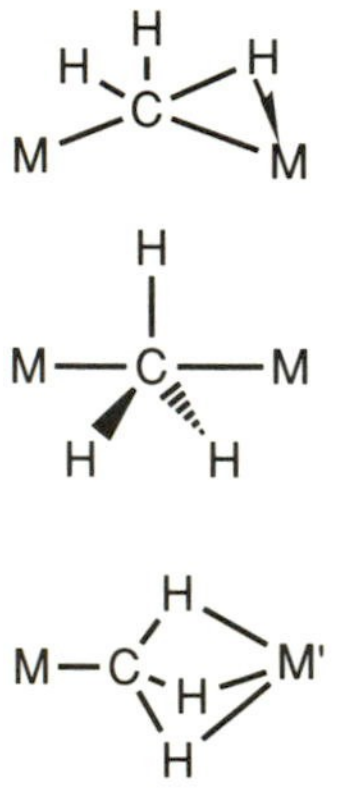

Terminal	Bridging

The existence of square-planar carbon is a point of topical debate. Examples have now been found: in the Zr complex below the angle sum around C is 360°.

Bonding to electron-deficient metal centres.

α-agostic   β-agostic η^2-benzyl

Metallacycles:

D = electron donor (O, N, P)

3.2 Synthesis

By alkylation of metal halides

This is the most widely applicable synthetic method, both for homoleptic metal alkyls and for complexes stabilized by donor ligands.

'Homoleptic' is the term applied to complexes with one uniform type of ligands, such as MMe_4 or $M(PMe_3)_4$, as distinguished from 'binary' compounds which consist of only two elements, e.g. MCl_4.

$$TiCl_4 \ + \ 2 \ Mg(CH_2SiMe_3)_2 \ \xrightarrow[\ -\ MgCl_2\ ]{-78\ °C\ to\ RT} \ Ti(CH_2SiMe_3)_4$$

Yellow, tetrahedral, diamagnetic (8 VE)

$$WCl_6 \ + \ 3 \ Al_2Me_6 \ \xrightarrow[-\ Me_2AlCl]{} \ WMe_6 \ \xrightarrow{LiMe} \ Li_2[WMe_8]$$

Colourless, octahedral, diamagnetic (12 VE)

Complex anion, 8-coordinate

$$Mo_2(OAc)_4 \ + \ 8 \ MeLi \ \longrightarrow \ [Li(THF)^+]_4 \left[\begin{array}{c} Me \\ Mo \equiv\!\equiv Mo \\ Me \end{array} \right]^{2-}$$

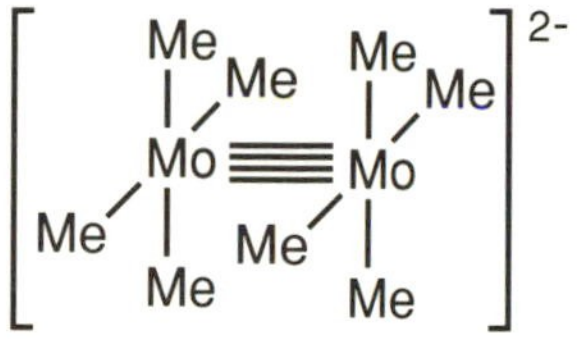

Anionic dimer, Mo–Mo quadruple bond

$$Cp_2MCl_2 \ + \ 2 \ RLi \ \xrightarrow{-\ LiCl} $$

$$R = CH_3, \ C_6H_5, \ C\equiv CR', \ \text{etc.}$$

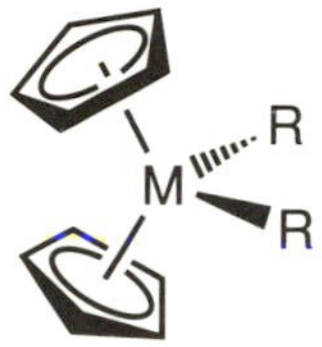

M = Ti, Zr, Hf

$$trans\text{–}NiCl_2L_2 \ + \ RMgCl \ \longrightarrow \ Cl\text{–}Ni\text{–}R$$

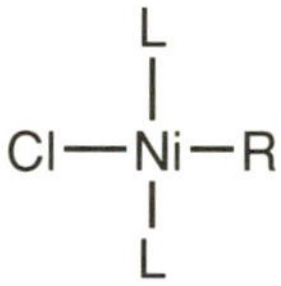

L = PMe_3
R = CH_2Bu^t

$$(R_3P)_2PtCl_2 \ + \ Li\!\!\diagdown\!\!\diagup\!\!\diagdown\!\!Li \ \longrightarrow$$

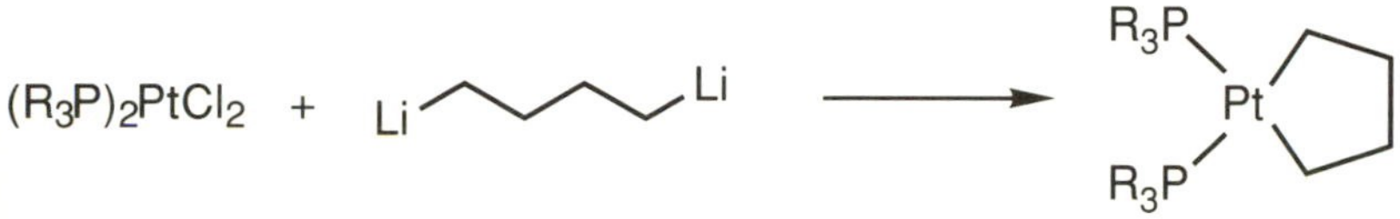

Alkylations may involve a change in the metal oxidation state:

$$Rh^{II}_2(OAc)_4 \ \xrightarrow[PMe_3]{Me_2Mg} \ Rh^{III}Me_3(PMe_3)_3$$

The reaction of $CrCl_3$ with PhMgCl in THF gave the first metal–aryl complex, $CrPh_3(THF)_3$ (*F. Hein, 1919*). The compound is stable since d^3 ions such as Cr^{III} are substitutionally inert.

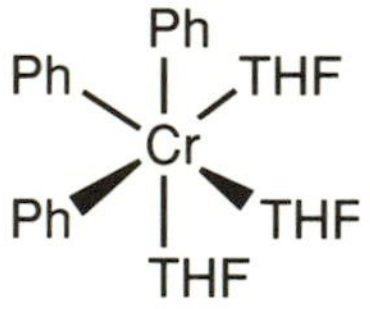

$$CrCl_3(THF)_3 \xrightarrow{\text{LiR}} CrR_4$$

Red-brown crystals, tetrahedral, paramagnetic (d^2)

$R = Me, Pr^i, Bu^t, CH_2Bu^t, .$

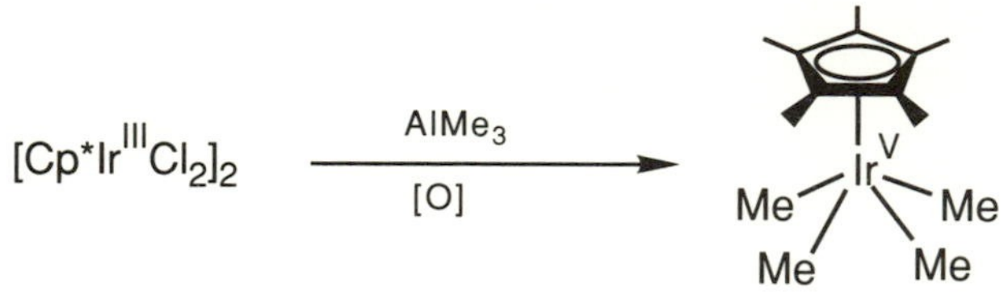

$$[Cp^*Ir^{III}Cl_2]_2 \xrightarrow[\text{[O]}]{\text{AlMe}_3}$$

Suitable choice of the alkylating agent allows the synthesis of alkyl halides and related mixed-ligand complexes. The alkylating strength of main group alkyls decreases with increasing covalent character of the metal–carbon bond: $LiR > RMgX > AlR_3 \geq ZnR_2$.

$$TiCl_4 \xrightarrow{\text{AlMe}_2\text{Cl}} MeTiCl_3 \xrightarrow{\text{ZnMe}_2} Me_2TiCl_2$$

$$TiCl_4 \xrightarrow{\text{MeLi}} TiMe_4$$

By oxidative addition

Low-valent metal complexes, most notably those of Ir^I, Ni^0, Pd^0 and Pt^0 stabilized by phosphines, undergo oxidative additions with alkyl, alkenyl, benzyl, and aryl halides. The reaction proceeds most readily with coordinatively unsaturated complexes (e.g. square–planar) and those containing labile ligands (e.g. complexes of ethylene or readily dissociating phosphines).

$$Ir^ICl(CO)(PPh_3)_2 + MeI \longrightarrow Ir^{III}Me(I)(Cl)(CO)(PPh_3)_2$$

Vaska's complex

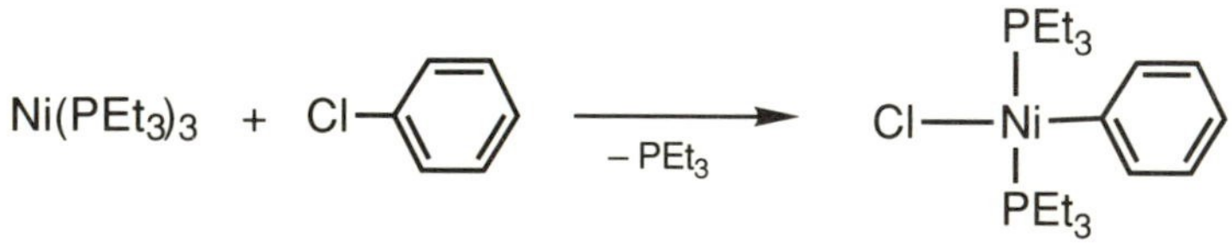

$$Ni(PEt_3)_3 + Cl-\bigcirc \xrightarrow{-PEt_3}$$

If the number of ligands that can be accommodated within the coordination sphere of the metal is restricted, ionic complexes may result:

$$CpCo(PMe_3)_2 \; + \; MeI \; \longrightarrow \; \left[\begin{array}{c} \text{Me}_3\text{P}-\overset{\displaystyle \bigcirc}{\underset{\displaystyle \text{Me}_3\text{P}}{\text{Co}}}-\text{Me} \end{array} \right]^{+} \quad I^{-}$$

Depending on the nature of the halide and the solvent, one-electron pathways involving electron transfer to Ni^{II} to give Ni^{I} species and radicals may compete with two-electron reactions. Reduced metal complexes may result, accompanied by radical coupling of the aryl group.

Oxidative addition is strongly favoured if the metal possesses a high-energy electron pair which can be attacked by the electrophile. This is the case for CpCo(PMe$_3$)$_2$: PMe$_3$ is strongly electron-donating and increases the electron density of the metal centre. The complex is a strong **metal base** and undergoes oxidative addition reactions very much more readily than CpCo(CO)$_2$.

$$NiL_4 \; + \; Ar{-}X \quad \underset{\substack{\text{polar solvent} \\ X = I}}{\overset{\substack{\text{non-polar solvent} \\ X = Cl}}{\Bigg\langle}} \quad \begin{array}{c} X{-}\overset{\displaystyle L}{\underset{\displaystyle L}{Ni}}{-}Ar \\[2em] L{-}\overset{X}{\underset{X}{Ni{\diagdown}Ni}}{-}L \quad + \quad Ar{-}Ar \end{array}$$

Low-valent complexes of early transition metals have a high tendency to obtain the highest possible oxidation state. They react with alkenes and alkynes frequently with formation of metallacycles; the reaction amounts to an oxidative addition. The metallacycle formation with alkenes is reversible.

$$Cp_2ZrPh_2 \; \xrightarrow[-\,Ph{-}Ph]{hv} \; \left[Cp_2Zr^{II} \right] \; \underset{Ph\!\equiv\!\!\equiv\!Ph}{\overset{C_2H_4}{\Bigg\langle}}$$

$$\left[Cp_2Zr^{II} \overset{\diagup\!\!\diagdown}{\underset{\diagdown\!\!\diagup}{}} \right] \; \rightleftharpoons \; Cp_2Zr^{IV}\!\!\diagup\!\!\diagdown$$

A special case of M–C bond formation via oxidative addition is the reaction with C–H bonds. This is facile with electron-rich platinum metals, particularly if the C–H bonds are 'activated'. Less basic complexes are generally unreactive.

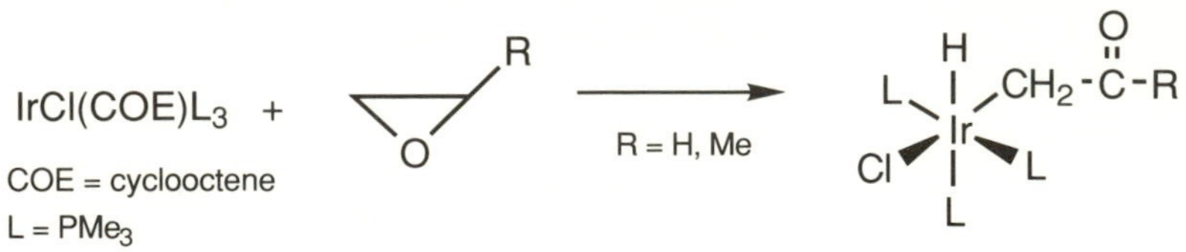

$$IrCl(COE)L_3 +$$

COE = cyclooctene
L = PMe₃

(epoxide, R = H, Me) → Ir complex

R = H, Me

Metal carbonyl anions are strongly nucleophilic and react readily with electrophiles, such as alkyl halides. The oxidation state of the metal changes by two units. A wide range of alkyl and acyl complexes is accessible by this route, including those containing β-hydrogens. The process corresponds to a nucleophilic substitution on carbon.

$$CpFe(CO)_2^-\ Na^+ + Et–I \longrightarrow CpFe(CO)_2Et + NaI$$

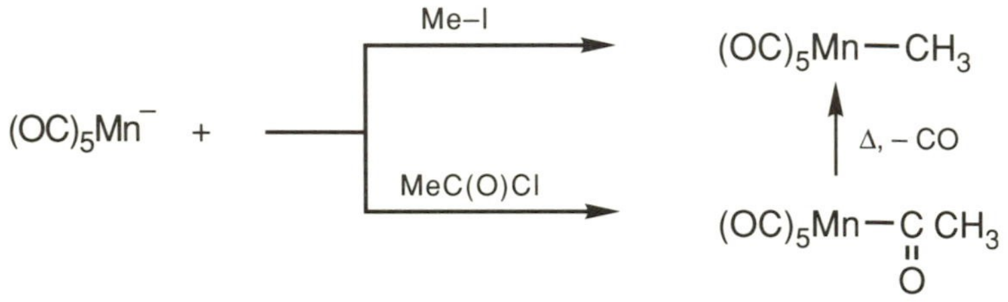

$$(OC)_5Mn^- +$$

Me–I → $(OC)_5Mn–CH_3$

MeC(O)Cl → $(OC)_5Mn–\underset{\underset{O}{\|}}{C}CH_3$

Δ, – CO

By addition reactions to metal hydrides

Alkenes and alkynes insert readily into M–H bonds. This reaction is the reverse of the β-hydride elimination reaction discussed above. Reactions of this kind – e.g. the formation of different isomers of thermally unstable cobalt alkyls – are important steps in catalytic reactions (see. p. 64). The insertion of alkenes into M–H bonds is frequently reversible; some β-H eliminations occur at low temperature, some require forcing conditions, depending on the M–C bond strength.

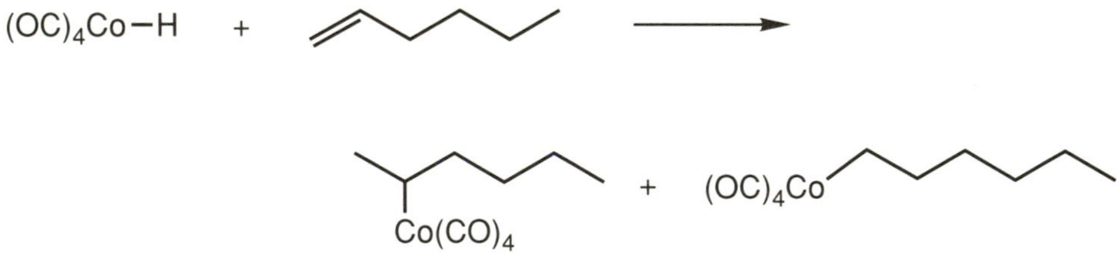

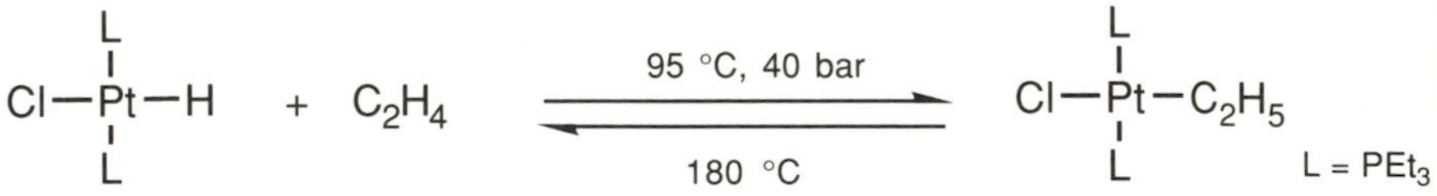

$$(OC)_4Co–H +$$

$$\text{(branched alkyl)}–Co(CO)_4 + (OC)_4Co–\text{(linear alkyl)}$$

$$\underset{L}{\overset{L}{Cl–Pt–H}} + C_2H_4 \underset{180\ °C}{\overset{95\ °C,\ 40\ bar}{\rightleftharpoons}} \underset{L}{\overset{L}{Cl–Pt–C_2H_5}}$$

L = PEt₃

The protonation of metal alkene and alkyne complexes similarly results in metal alkyl (alkenyls) and may involve a metal hydride intermediate:

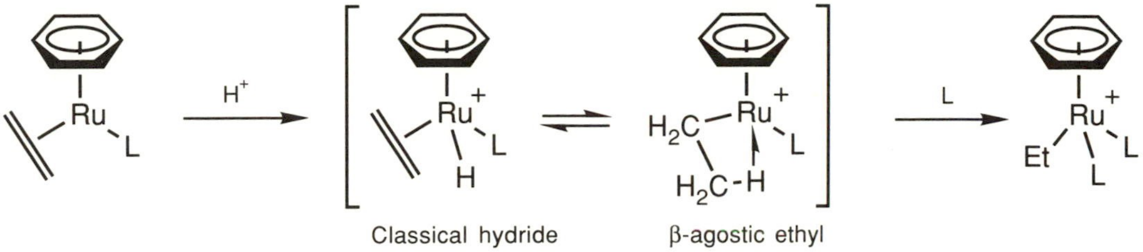

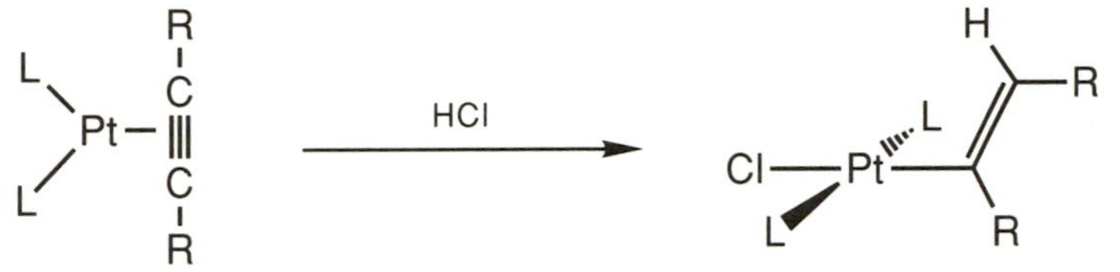

Classical hydride   β-agostic ethyl

Diazomethane reacts with acidic metal hydrides to give methyl complexes, a reaction similar to the formation of methyl esters from carboxylic acids:

$$CpMo(CO)_3H \ + \ CH_2N_2 \longrightarrow CpMo(CO)_3CH_3 \ + \ N_2$$

By nucleophilic attack on coordinated ligands

Polar unsaturated ligands, particularly CO, are susceptible to nucleophilic attack. This reaction is frequently observed with cationic complexes.

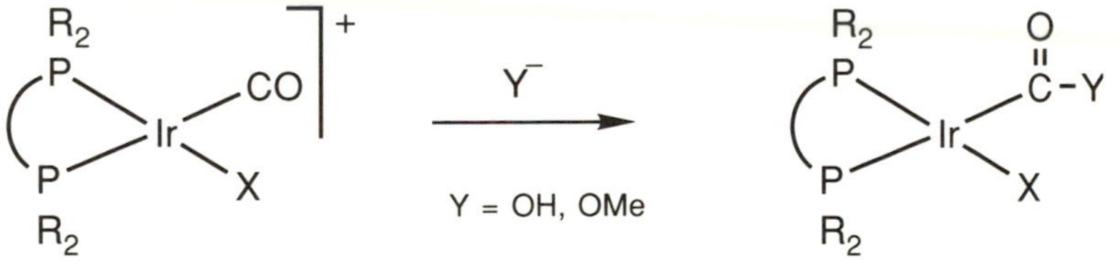

$$Fe(CO)_5 \ + \ LiR \longrightarrow Li^+ \left[(OC)_4Fe-\overset{\displaystyle O}{\underset{\displaystyle \|}{C}}-R \right]^-$$

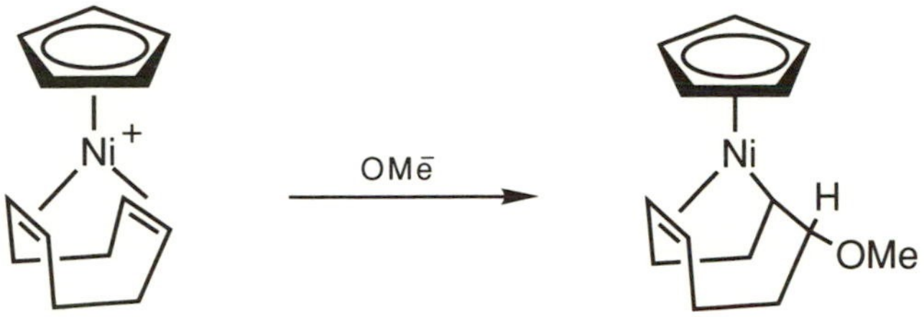

By *ortho*-metallation

Ortho-metallation involves the cleavage of *ortho*-C–H bonds of aryl groups by the metal, with formation of M–C bonds (see C–H activation, p. 72). The reaction is frequently encountered with aryl groups bound to coordinated donor-atoms, such as N or P, and with electron-rich metals in low oxidation states which are able to undergo oxidative additions.

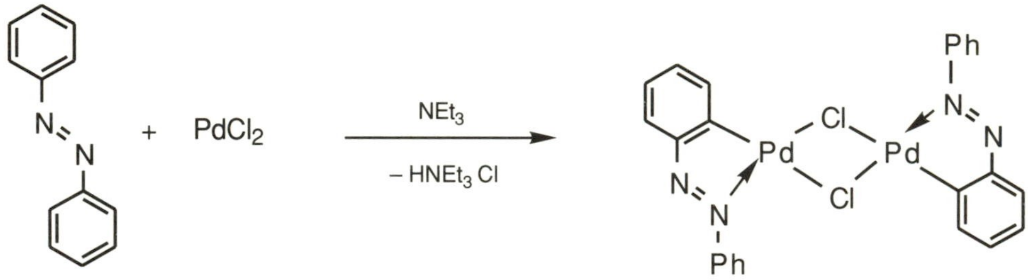

Ortho-metallation is not uncommon for PPh$_3$ complexes of noble metals and is assisted by steric crowding.

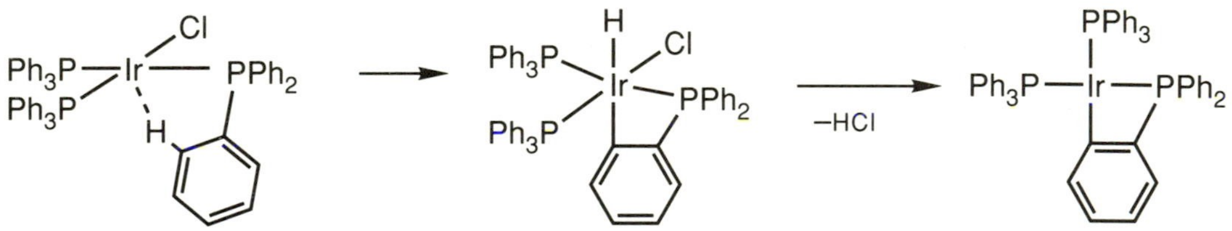

3.3 Metal alkyls and the 18–electron rule: The importance of steric stabilization

The structures and stability of most complexes encountered until now, such as the transition metal carbonyls and many other π-complexes, are conveniently explained by the 18-electron rule (see. p. 2) which predicts stability for a complex in which the metal is able to attain the electron configuration of a noble gas. CO is sterically relatively undemanding, and the metal is usually able to accommodate the required number of ligands to satisfy its electron requirements.

Most metal alkyl complexes containing donor ligands also follow the 18-electron rule (16 electrons for square–planar d^8 systems: Rh^I, Ir^I, Ni^{II}, Pd^{II}, and Pt^{II}). An early example is Pope's Me$_3$PtI: the 18-electron rule is fulfilled if the iodide ligands act as triply-bridging five-electron donors (one σ-bound with two electron pairs as donors – rather like the cyclopentadienyl ligand in CpPtMe$_3$) and the molecule adopts a tetrameric cubic structure.

	Covalent counting formalism		Ionic counting formalism	
	Pt^0	d^{10}	Pt^{IV}	d^6
	3 Me	3	3 Me$^-$	6
	I	5	I$^-$	6
		18 VE		18 VE

Other typical examples are:

d^7 + 5×2 + 1 = 18 d^6 + 3×2 + 5 +1 = 18 d^8 + 4×2 + 2×1 = 18 d^{10} + 2×2 + 2×1 = 16

On the other hand, there are numerous complexes which are thermally quite stable but cannot be explained on the basis of the 18-electron rule. These compounds are often electronically highly unsaturated and owe their existence to the *kinetic stabilization* afforded by sterically demanding ligands. These ligands are often too bulky to allow the metal to bind enough donor ligands to achieve a higher electron count; on the other hand, they provide an efficient 'umbrella' protecting the metal. Examples are found mainly among early transition and lanthanide metals in high oxidation states.

The importance of kinetic shielding: steric versus electronic saturation

TiR$_3$ (Ti^{III}, d^1, 7 VE): alkyl ligand too bulky to form TiR$_4$

TiR$_4$ (Ti^{IV}, d^0, 8 VE) preferred with smaller ligands

In both these cases, other metals, such as vanadium and chromium, form very similar tri- and tetraalkyls, irrespective of their different electron configurations.

The ligand –C(SiMe$_3$)$_3$ is even bulkier. Manganese(II) dialkyls are usually polymeric, with tetrahedrally coordinated Mn and structures similar to

MgMe$_2$ or BeMe$_2$, in accord with the ionic nature of Mn^{II} (d^5). By contrast, Mn[C(SiMe$_3$)$_3$]$_2$ is a two-coordinate monomer. Similarly, zinc dimethyl, a linear, monomeric, volatile compound, is highly pyrophoric and instantly hydrolysed, whereas Zn[C(SiMe$_3$)$_3$]$_2$ can be boiled in aqueous HCl without decomposition!

Polymeric, tetrahedral

Monomeric, linear

H$_3$C—Zn—CH$_3$
Monomeric, linear
highly pyrophoric

(Me$_3$Si)$_3$C—Zn—C(SiMe$_3$)$_3$
Stable to air and acids

A particularly striking example of the overriding importance of steric factors in determining the geometry and the oxidation state of the metal are 1-norbornyl complexes, M(1-norb)$_4$: a series of these exists with M = Ti, V, Cr, and even tetravalent Mn, Fe and Co.

M = Ti, V, Cr, Mo, Mn, Fe,

The Co^{IV} compound is a rare example of a **low-spin tetrahedral** complex – an indication of the strong ligand field exerted by powerful σ–donors such as alkyl ligands. It can be reduced to [Co^{III}(norb)$_4$]$^-$ and oxidised to the unusual Co^V complex [Co(norb)$_4$]$^+$. All these complexes are tetrahedral.

$$[Li(THF)_4]^+ \ [Co(1\text{-norb})_4]^- \underset{Li,\ THF}{\overset{CoCl_2}{\rightleftharpoons}} Co^{IV}(1\text{-norb})_4 \xrightarrow{Ag^+} [Co^V(1\text{-norb})_4]^+$$

Green, 10 VE

d^6 (μ_{eff} = 3.18 B.M.)

Red-brown, 9 VE

low–spin d^5 (μ_{eff} = 2.0 B.M.)

Brown, 8 VE

d^4

Although these examples demonstrate how important steric factors can be in controlling the maximum attainable coordination number, electron-deficient complexes nevertheless attempt to increase their electron count by whatever means available. Benzyl ligands, for example, possess a π-system (the phenyl ring) which is able to act as a weak electron donor towards the metal centre. A typical example is tetrabenzylzirconium: the *ipso*-carbon of the phenyl ring interacts with Zr; the benzyl ligand is, in effect, η^2-*coordinated*. As a result, the Zr–CH$_2$–C angle is considerably more acute than expected for an sp^3-carbon. By contrast, Sn(CH$_2$Ph)$_4$ shows none of these distortions and possesses normal angles: as a main group element without accessible unoccupied *d*-orbitals, its benzyl ligands are η^1-bonded.

η^1–benzyl η^2–benzyl η^3–benzyl

The η^2-bonding mode in Zr(CH$_2$Ph)$_4$ should be distinguished from the η^3-*allylic bonding* sometimes encountered with benzyls in electron–rich systems (see *Organometallics 2*, Chapter 5). A typical example is CpMo(CO)$_2$(η^3-CH$_2$Ph): the metal is coordinated to three carbons in π-fashion, and the aromatic character of the phenyl ring is disrupted (note the bond length alternation in the six–membered ring in the Mo example!).

Donor interactions increasing the effective electron count of the metal centre do not have to be so obvious. Heteroatoms, such as O and N, are able to interact with the metal to which they are bound via their 'non-bonding' electron pairs. This is common in imido and oxo complexes. Indicative for such interactions are the short metal–N and metal–O bonds and the near-linear M–O–M arrangement in binuclear complexes.

16/18 VE   11/13/15 VE?   14, 16 or 18 VE?   14/18 VE ?

Many of these compounds are very stable; MeReO$_3$ for example is air-stable and even water soluble; it is a good catalyst for the epoxidation of olefins with H$_2$O$_2$ and, attached to an SiO$_2$ support, for olefin metathesis.

The extent of the steric saturation of a complex depends both on the alkyl ligand and on the radius of the metal centre. For example, TiMe$_4$ can bind one chelating phosphine ligand such as dmpe, whereas ZrMe$_4$ can accommodate two. Both are 'coordinatively saturated' and stable at ambient temperature.

Sterically hindered *n*-donor ligands, notably phosphines, can be as effective as alkyl ligands in restricting the coordination number and the electron count of the metal centre. Not only does the ligand environment dictate the coordination geometry and the degree of electronic unsaturation but it can also determine other properties such as magnetism:

12 VE

16VE

Dimer, diamagnetic, d^4, 14 VE
Cr–Cr quadruple bond

Distorted tetrahedral:
d^4 high spin, μ_{eff} = 4.9 B.M.
(12 VE)

Octahedral: d^4 low spin,
μ_{eff} = 2.7 B.M. (16 VE)

3.4 Reactivity of metal alkyls

Most transition metal alkyls are highly reactive. The most important reaction types are:

☞ *reactions with M–C bond cleavage: M–C bond hydrogenolysis; reactions with electrophiles. Such as CO_2, H^+, or I_2;*

☞ *M–C insertion reactions with unsaturated organic molecules. Such as CO, isocyanides, alkenes, or alkynes;*

☞ *reductive elimination reactions:, for example coupling of alkyl ligands with nucleophiles.*

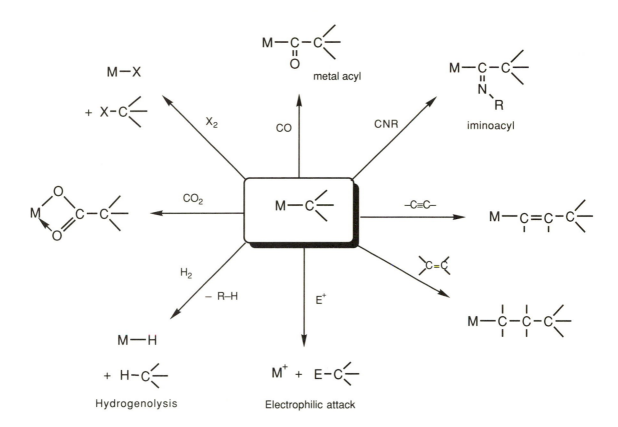

M–C bond cleavage reactions

Reactions with electrophiles
Generally, metal–carbon σ-bonds react readily with electrophiles such as HX
to give alkanes and complexes L_nMX. The reaction of water or aqueous
mineral acids can be violent with electron-deficient alkyls of early transition
metals but goes smoothly with more electron-rich complexes. Alcohols,
phenols and ammonium salts may be used, for example:

$$Zr(CH_2Ph)_4 \xrightarrow[-\text{toluene}]{ArOH} Zr(CH_2Ph)_3(OAr) \xrightarrow[-\text{toluene}]{ArOH} Zr(CH_2Ph)_2(OAr)_2$$

$$Cp_2TiMe_2 \xrightarrow[-CH_4]{NH_4^+ PF_6^-} \left[Cp_2Ti\begin{smallmatrix} Me \\ \\ NH_3 \end{smallmatrix} \right]^+ PF_6^-$$

$$NiMe_2(PMe_3)_2 \xrightarrow[\substack{-CH_4 \\ -PMe_3}]{H_2O}$$

Metal electrophiles, such as $HgCl_2$, react with transition metal alkyls usually with alkyl transfer to mercury. Two reaction pathways have been observed with stable alkyl complexes, e.g. of iron. In protic solvents and if R is primary alkyl, generation of a mercury alkyl is favoured, whereas in aprotic solvents and if R is a secondary or tertiary alkyl, alkyl halides and complexes with mercury–iron bonds are formed. Alkyl transfer also takes place with other main group halides, such as thallium trichloride.

R = primary alkyl

$$R–HgCl \ + \ FeCl(CO)_2Cp$$

$HgCl_2$

R = *sec. or tert. alkyl*

$$R–Cl \ + \ ClHg–Fe(CO)_2Cp$$

Halogens are frequently used to cleave metal–carbon bonds. The reaction can be used to obtain functionalised alkyl derivatives via transition metal mediated synthesis.

$$Cp(OC)_2Fe–R \ + \ Br_2 \longrightarrow Cp(OC)_2Fe–Br \ + \ R–Br$$

The mechanism of the reaction in aprotic solvents involves electrophilic attack by Br^+ on the metal. If $R = CD_2CH_2Ph$, both possible bromoalkane isomers are produced, via a cyclic intermediate:

$$Cp(CO)_2Fe–CD_2CH_2Ph \ + \ Br_2 \longrightarrow \left[Cp(OC)_2Fe \begin{smallmatrix} Br \\ CD_2CH_2Ph \end{smallmatrix} \right]^+ Br^-$$

$$\longrightarrow \left[CpFe(CO)_2Br + \underset{H_2C–CD_2}{\overset{+}{\bigcirc}} \right] Br^- \longrightarrow \begin{matrix} PhCH_2CD_2Br \\ PhCD_2CH_2Br \end{matrix}$$

Carbon dioxide can give mono- and bidentate insertion products. Hydrolysis of the resulting complexes can give the corresponding carboxylic acids. The reaction is similar to the reactions of Grignard reagents.

$$Cp_2TiMe_2 \; + \; 2 \; CO_2 \longrightarrow Cp_2Ti(O_2CMe)_2$$

$$L_3Rh\text{--}R \; + \; CO_2 \longrightarrow L_3Rh\text{--}O\text{-}\underset{\underset{O}{\|}}{C}\text{-}R$$

R = Me, Ph
L = PPh₃

Carbon disulfide inserts in an analogous fashion to CO_2. Sulfur dioxide is a comparatively strong electrophile and gives insertion products which may be S-bound or O-bound. The reactions are usually conducted in liquid SO_2 at low temperature.

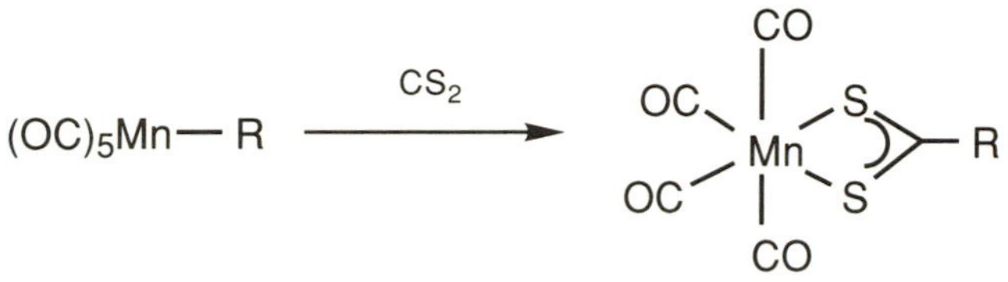

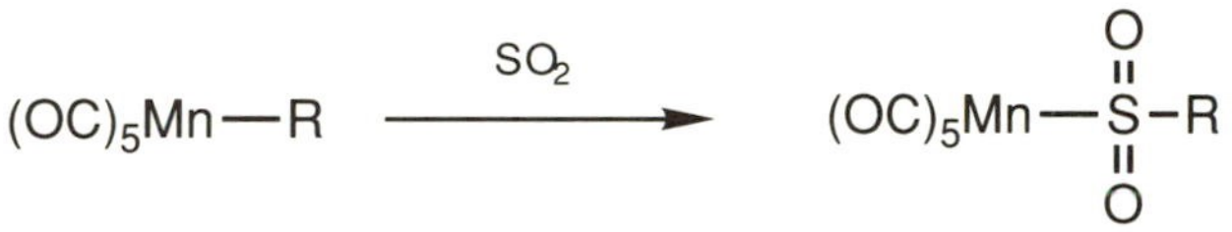

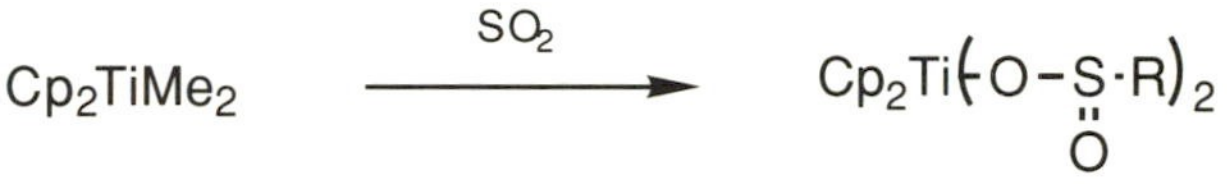

Insertions of NO

NO inserts into M–C bonds in a manner reminiscent of CO. However, the unpaired electron of NO has to be accommodated. Diamagnetic metal alkyls react with NO under double insertion to give a metallacycle:

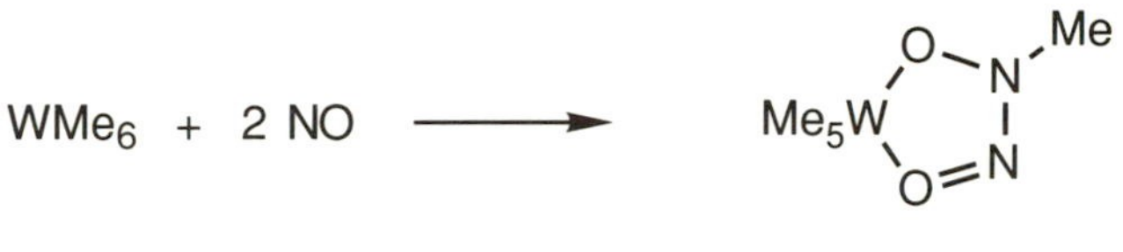

With paramagnetic alkyls oxidation usually results, and in the presence of olefins the transfer of nitrene to the C=C double bond has been observed:

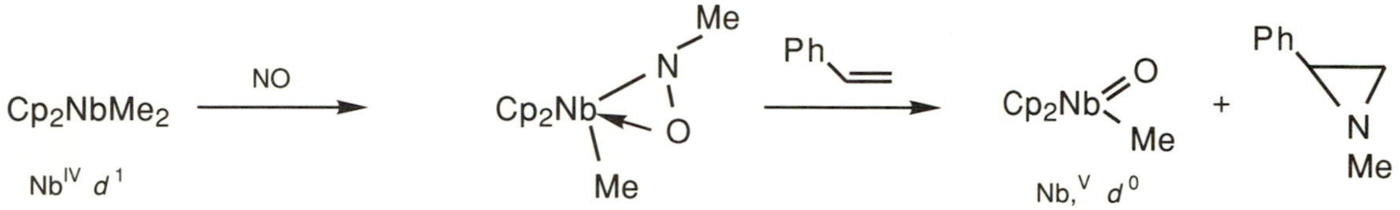

Reactions with H_2

The reaction of dihydrogen with metal alkyls is of widespread importance. It occurs in several catalytic reactions (see hydrogenation, hydroformylation p. 64) and is often a clean synthetic method for the preparation of metal hydrides.

$$Cp^*{}_2ZrMe_2 \quad \xrightarrow{H_2,\ pressure} \quad Cp^*{}_2ZrH_2 \quad + \quad 2\ CH_4$$

$$(OC)_4Co{-}R \quad \xrightarrow{H_2,\ pressure} \quad (OC)_4Co{-}H \quad + \quad R{-}H$$

Insertion of CO into M–C bonds

The insertion of CO into M–C bonds to give acyl complexes is one of the most widely applicable and industrially most important reactions of transition metal alkyls. For example, if $CH_3Mn(CO)_5$ is treated with CO under pressure, $CH_3C(O)Mn(CO)_5$ results. This could be formed either by attack of free CO on the metal–alkyl bond, or by migration of the alkyl ligand to a coordinated CO. The mechanism becomes apparent if ^{13}C-labelled CO is used as the incoming ligand: the acyl group is derived from previously coordinated CO, while the labelled ligand occupies a position *cis* to the acyl group. This is consistent with an intramolecular migration of the nucleophilic CH_3 ligand on to the electrophilic carbon of a coordinated CO ligand; the coordination site thus vacated by the methyl ligand is then occupied by [13]CO.

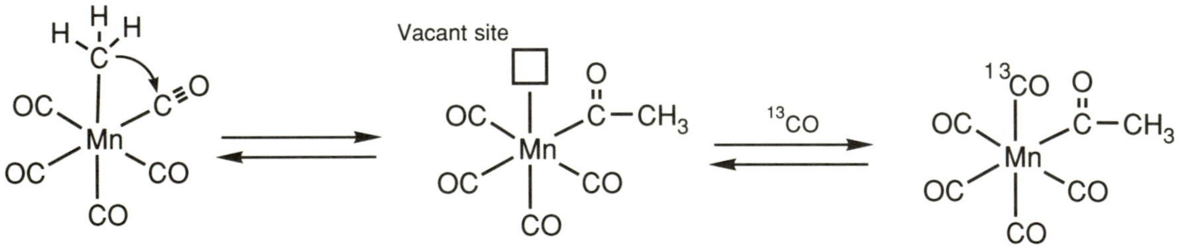

The reaction is reversible: if $CH_3C(O)Mn(CO)_5$ is heated under vacuum, a CO ligand – again *cis* to the acyl ligand – dissociates and allows the methyl group to migrate from the acyl group to the metal to generate $CH_3Mn(CO)_5$. This process highlights several important reaction principles:

☞ The 'insertion' of CO into the M–C bond consists of a **migration of the alkyl ligand** to a coordinated CO ligand.

☞ Migration represents an **intramolecular nucleophilic attack** by alkyl on a coordinated unsaturated electrophile.

☞ The alkyl migration step is **reversible.**

The validity of these statements is not confined to CO insertion: the principle of *cis*-migration applies equally well to the 'insertion' of alkenes, alkynes, CO_2, etc. The incoming ligand need not be identical to the one that participates in the alkyl migration step; with $CH_3Mn(CO)_5$ the equilibrium is equally displaced in favour of acyl formation if, for example, PPh_3 is present, giving rise to $CH_3C(O)Mn(CO)_4(PPh_3)$.

Lewis acids (BF_3, $AlCl_3$) greatly accelerate the rate of alkyl migration, first by coordinating to a CO ligand and increasing its polarity, and secondly by weakly occupying the vacant coordination site, thereby stabilizing the intermediate before the incoming ligand takes its place.

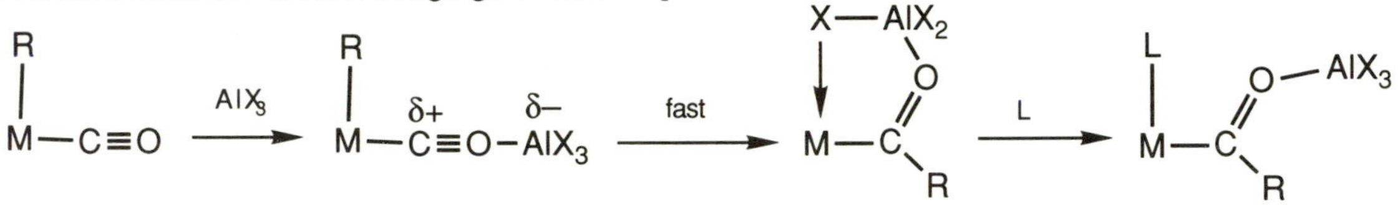

AlX3
δ+
δ−
fast
X—AlX2
L
O—AlX3

If chiral alkyl ligands are used, it can be shown that the migration step proceeds with *retention of configuration:*

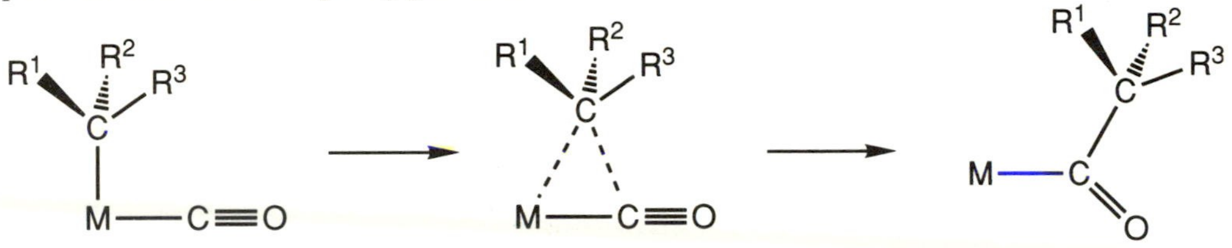

Many alkyl complexes do not contain coordinated CO but nevertheless react readily if exposed to CO to give acyl complexes. However, the same principle of *cis*–migration applies and has been demonstrated in the reaction of CO with Cp_2ZrPh_2:

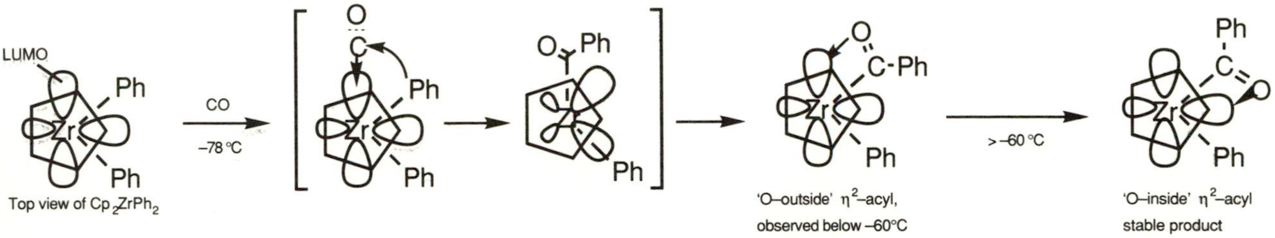

LUMO
CO
−78 °C
>−60 °C
Top view of Cp2ZrPh2
'O–outside' η2–acyl, observed below −60°C
'O–inside' η2–acyl stable product

The incoming CO ligand forms a very labile complex by interacting with the LUMO on zirconium. Alkyl migration is fast and leads to an acyl product as expected. The 16 VE η^1-acyl complex which may be formed initially is stabilized by the donor capacity of the acyl–oxygen to give a complex which contains an η^2-bonded acyl ligand: a low-temperature isomer ('O-outside') can be identified below –60 °C and rearranges on warming to give the stable product, with an 'O-inside' η^2-acyl ligand.

η^2-Acyls and η^2-iminoacyls are frequently encountered in electron–deficient complexes, particularly of early transition metals, lanthanides and actinides, although examples for late transition metals such as Fe and Ni are also known.

Double carbonylation

If CO so readily inserts into M–C bonds, can this process be repeated to give α-keto acyls or even poly-ketones? This is not the case: the insertion of CO into a metal–acyl bond is thermodynamically unfavourable and is not observed. $CH_3C(O)C(O)Mn(CO)_5$ can, however, be prepared from $Mn(CO)_5^-$ and $CH_3C(O)C(O)Cl$; it is thermally unstable and readily eliminates CO to give $CH_3C(O)Mn(CO)_5$.

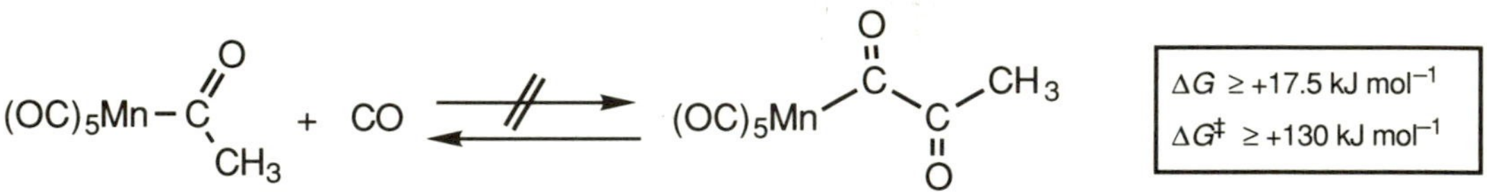

Nevertheless, 'double carbonylation' processes are occasionally found. The carbonylation of iodobenzene in the presence of palladium catalysts for example gives α-keto benzoic amides:

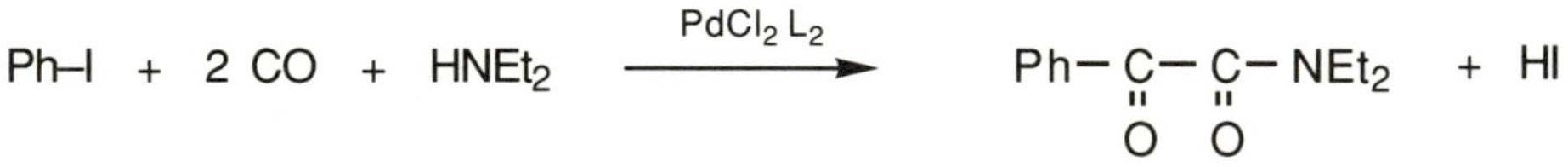

Investigation of the mechanism shows, however, that the product is the result of the reductive coupling of two different CO-containing units, C(O)Ph and C(O)NEt$_2$; the latter is formed by nucleophilic attack of HNEt$_2$ on coordinated CO in a cationic palladium carbonyl intermediate. PhC(O)NEt$_2$ is also produced via a side reaction; a Pd–CO–CO–Ph species is not involved in the cycle.

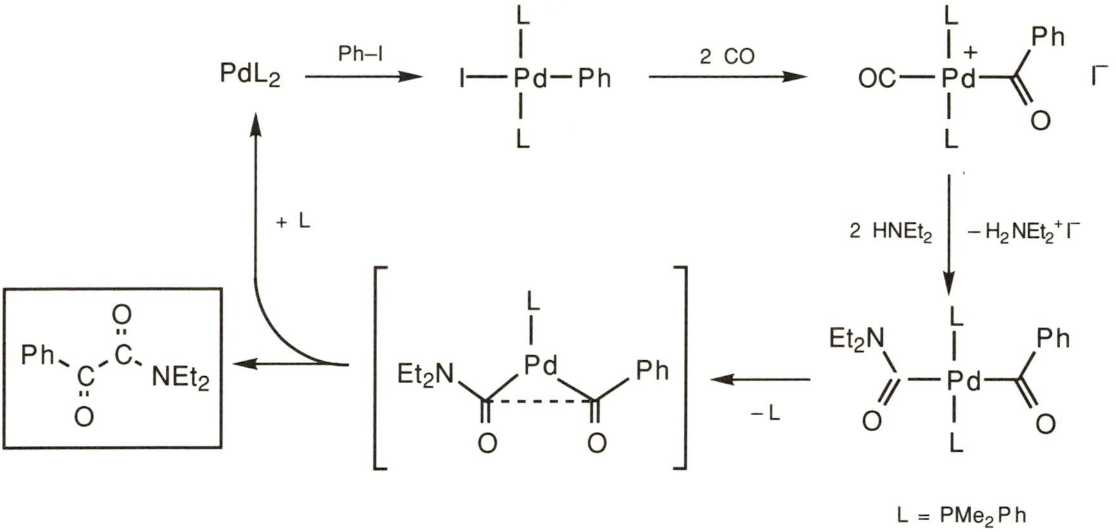

L = PMe₂Ph

The 'double carbonylation' of dialkyls of oxophilic metals, such as Zr or Th, has been shown to lead to C–C coupling reactions (see Fischer–Tropsch reaction, p. 33).

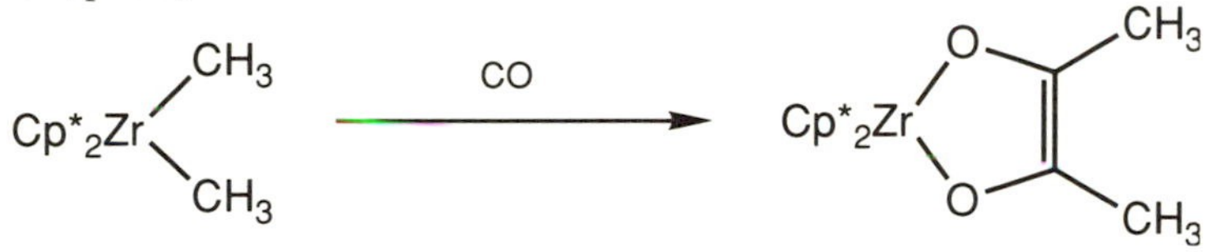

Insertion of isocyanides

Isocyanides are closely analogous to CO and generally undergo similar reactions (see p. 39); it is, however, less dependent on back-bonding for stabilisation of its complexes. Commonly used reagents are Bu^{t}NC and ArNC (Ar = e.g. *m*-xylyl). They react with metal alkyls and aryls to give iminoacyls by the mechanism described for CO. Because of the pronounced donor character of the iminoacyl nitrograph η^2 complexes are frequent. In contrast to CO, however, multiple insertions are no longer energetically unfavourable, and double and triple RNC insertion takes place; indeed, (allyl)Ni(O$_2$CCF$_3$) is highly effective in catalysing the polymerization of ArNC to give linear poly(iminoacyl). Some examples of the reactions of isocyanides with metal alkyls are:

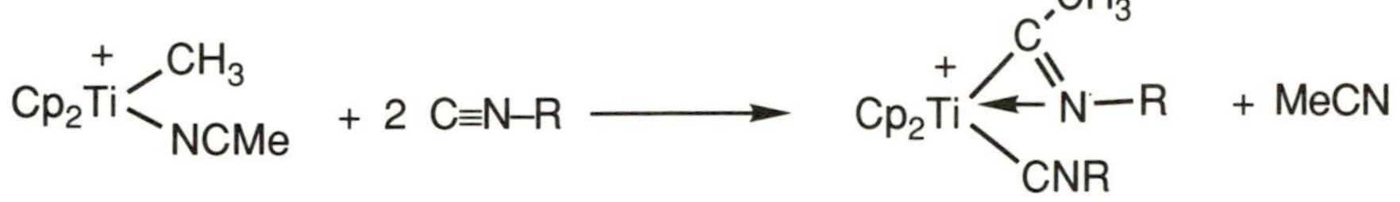

L = PMe$_3$

CO insertions in catalysis

Hydroformylation

Several industrial processes involve the insertion of CO into metal–alkyl bonds of catalytic intermediates. One of the best known is the **hydroformylation** of olefins, originally developed by Roelen (BASF) in 1938 using cobalt catalysts, like $Co_2(CO)_8$, and known as the '**oxo process**'. It operates at elevated temperatures and high CO/H_2 pressures; under such reducing conditions $HCo(CO)_4$ is formed which initiates the catalytic cycle (see p. 65).

The hydroformylation with cobalt carbonyls requires drastic conditions and high CO pressures to stabilize the $HCo(CO)_4$ intermediate. Efforts have been made to improve the conditions, for example by addition of PBu_3. Significantly more efficient catalysts are only obtained by replacing cobalt with rhodium; the hydrides and alkyls of second-row metals are less labile, and so much more active, that the high cost of rhodium is easily offset. The catalyst precursors are $HRh(CO)(PPh_3)_3$ or $RhCl(PPh_3)_3$ (*Wilkinson's catalyst*) in the presence of a large excess of PPh_3; this catalyst has now superseded the older cobalt system. The catalytic cycle is closely analogous to that outlined below for Co and is facilitated by the ready dissociation of PPh_3 to give 16VE species such as $HRh(CO)(PPh_3)_2$.

The reaction scheme below is simplified. The insertion of a terminal olefin into the metal hydride can occur in a Markovnikov and anti-Markovnikov sense:

Markovnikov product   anti-Markovnikov product

$(OC)_4Co-H$
18 VE

$+ CO \quad \Big\updownarrow \quad -CO$ ①

$(OC)_3Co^I-H$
16 VE

②

$(OC)_3Co\diagdown_H$ (with alkene R)
18 VE

③

$(OC)_3Co\diagup R$
16 VE

② | CO

$(OC)_3Co\diagup R$ with CO
18 VE

③

$(OC)_3Co^I-\overset{O}{\underset{\|}{C}}-R$
16 VE

H₂

④

Co^{III} complex: OC, OC, OC ligands, H, H, and $\overset{\|}{C}{=}O$ chain to R
18 VE

⑤

$R\diagdown\diagup\overset{H}{\underset{\|}{C}}{=}O$ (boxed product)

① Ligand dissociation – generation of unsaturated intermediate
② Substrate binding to vacant site
③ Alkyl (or hydride) migration (insertion step)
④ Oxidative addition
⑤ Reductive elimination

A mixture of primary and secondary cobalt alkyls results, which leads to linear and branched aldehydes. Since the aldehyde products are mainly converted to detergent alcohols which have to be linear, the branched isomer is undesirable. Rhodium catalysts, particularly in the presence of a large excess of triphenylphosphine, give predominantly linear products.

Dihydrogen can be replaced by protic reagents such as alcohols or amines to give a range of carbonylation products, including carboxylic acids, anhydrides, esters and amides. Many of these reactions were discovered by W. Reppe in the 1940's using cobalt salts.

Central box: **M–H + CO**

Ph—NCO ← $PhNO_2$ ← [M–H + CO]

[M–H + CO] → (allyl amine, NH_2) → pyrrolidinone (N–H, =O)

MeOOC—(chain)—COOMe ← MeOH ← [M–H + CO]

[M–H + CO] ↓ H_2O → $Me_3C-COOH$

[M–H + CO] → $H{\equiv}H$, H_2O → $\diagup\diagdown COOH$

Methanol carbonylation

Carbonylation reactions are not restricted to olefins. Cobalt salts in the presence of iodide convert methanol to a mixture of acetic acid and methyl acetate, another Reppe process:

$$MeOH + CO \xrightarrow{CoI_2} MeCOOH + MeCOOMe$$

Again this reaction is very much more efficient and selective if rhodium catalysts are used, and is the basis of the **Monsanto acetic acid process**. The reaction consists of two coupled cycles: an iodide cycle which converts the unreactive methanol into iodomethane, and the rhodium carbonylation cycle.

Higher alcohols are similarly carbonylated, although the reactivity decreases for MeOH > EtOH > PrOH. The mechanism has been elucidated mainly by kinetic and *in-situ* IR spectroscopic studies. An intermediate has been isolated, $[Rh_2(COMe)_2(CO)_2I_6]^{2-}$, which reacts with CO to give $[Rh(COMe)I_3(CO)_2]^-$. The Monsanto process now represents the major industrial route to acetic acid and is replacing older alternatives such as the oxidation of butane or the palladium/copper catalysed ethylene oxidation (Wacker process, see *Organometallics 2*, chapter 2).

Cobalt catalysts, in the presence of hydrogen, not only carbonylate alcohols to carboxylic acids but reduce the acids to alcohols, so the overall reaction leads to alcohol homologation.

A process for the production of acetic anhydride by carbonylation of methylacetate has recently been put into operation (Tennessee Eastman). Since methylacetate is itself made from methanol and CO, the synthesis relies purely on C_1 feedstocks.

$$CH_3OH + CO + 2H_2 \xrightarrow[180°C, 200 \text{ bar}]{Co_2(CO)_8, \, I^-} CH_3CH_2OH + H_2O$$

Insertions of alkenes and alkynes

One of the most important catalytic industrial processes relies on the facile insertion of alkenes into metal–alkyl bonds: the polymerization of olefins. In spite of intense efforts to find model systems which demonstrate this important reaction step in a controlled manner, and in contrast to the numerous examples of alkene insertions into metal–hydrogen bonds, surprisingly few well-characterized examples exist.

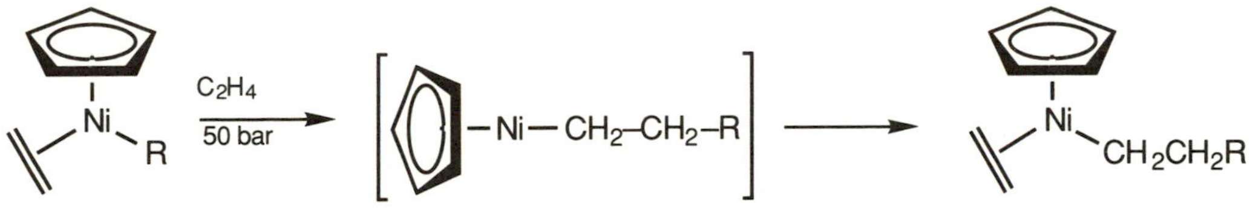

R = Ph, –CH$_2$cyclopropyl

Activated alkenes insert even into metal–acyl bonds; the reaction is of significance in view of the ethylene–CO copolymerization catalysed by nickel and palladium complexes.

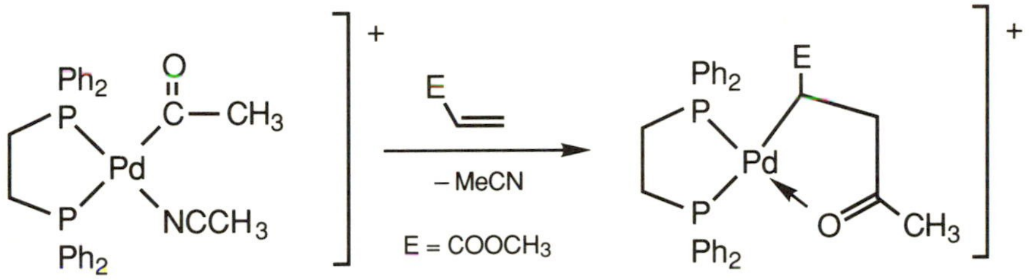

The mechanism follows, in principle, the pathway outlined for the 'insertion' of CO: the alkyl ligand and the coordinated alkene occupy positions *cis* to each other, to allow alkyl migration to occur:

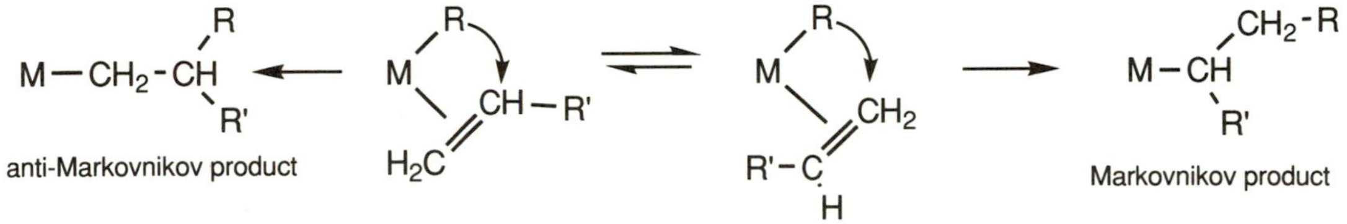

With terminal olefins, migration can be to the first (Markovnikov) or the second carbon of the double bond (anti-Markovnikov). Since M–C bonds to primary alkyls are sterically less hindered than to secondary alkyls, the anti-Markovnikov product is preferred.

The insertion of alkynes into M–C bonds in square–planar nickel(II) complexes has been shown to be preceded by ligand substitution.

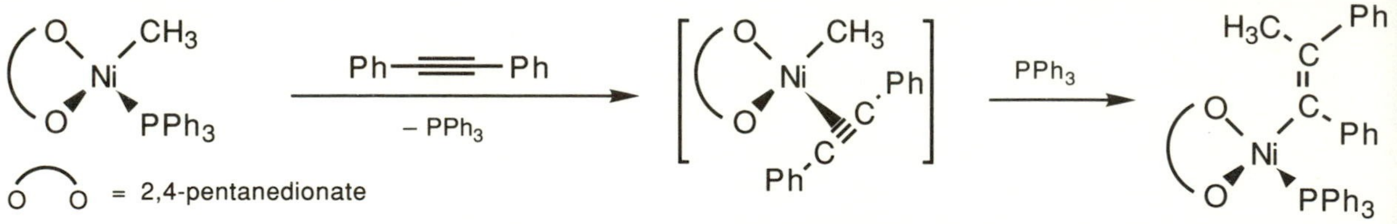

Multiple insertions of alkynes have been observed to give complexes free of coordinated dienyl ligands, depending on the electronic chraracteristics of the metal centre.

Related unsaturated compounds such as nitriles are also able to react with

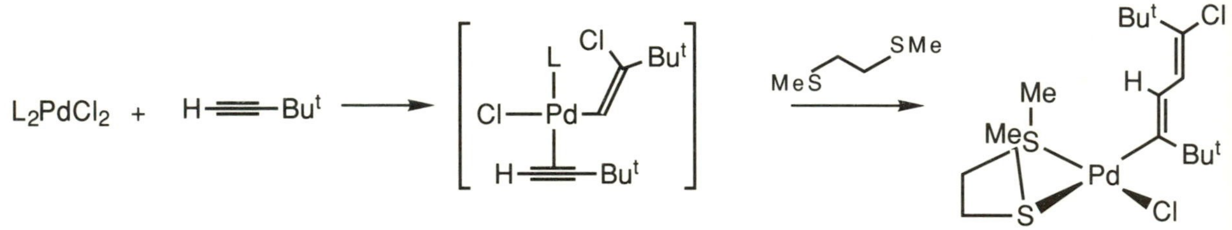

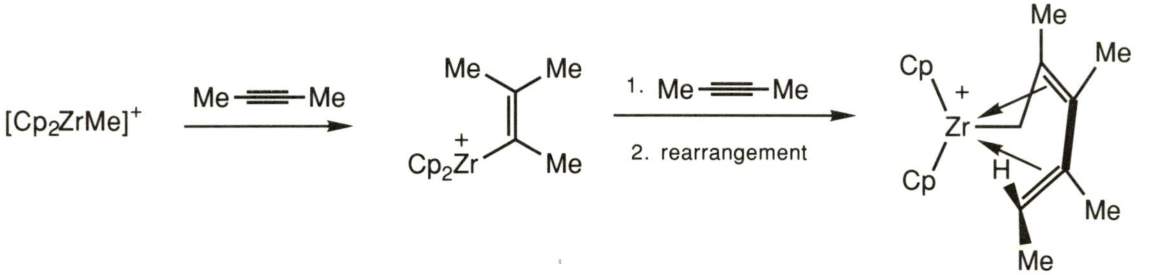

electron-deficient alkyls of early transition metals to give 'azavinylidene' complexes; typical is the linear M–N–C arrangement with pronounced M–N double bond character. In contrast to reactions with CO, coordinatively more saturated complexes such as Cp_2TiMe_2 do not react.

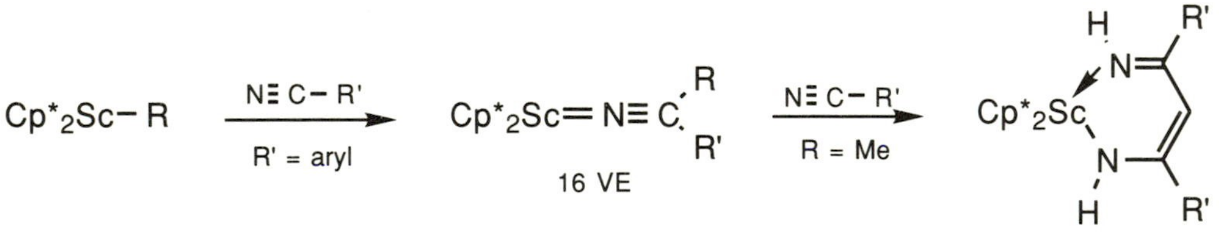

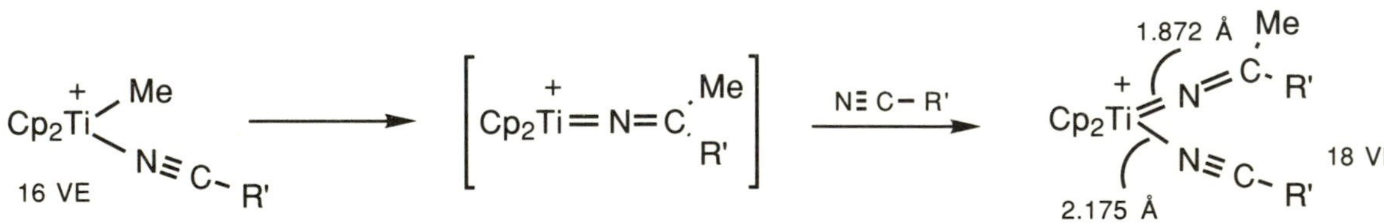

Ziegler–Natta polymerization

During experiments to synthesize long chain aluminium alkyls by treating aluminium triethyl with ethene under pressure ('*Aufbau reaction*'), K. Ziegler noted that transition metal halides have a dramatic effect on the course of the reaction. Whereas nickel salts led to the dimerization of ethene to butene, $TiCl_4$ catalysed its polymerization to give a relatively high melting linear polymer. G. Natta applied this catalyst to the stereospecific polymerization of propene. The polymerization of alkenes with metal halides activated by aluminium alkyls (**Ziegler–Natta catalysis**) is now one of the most important industrial processes.

$TiCl_4/AlEt_3$ mixtures as used by Ziegler react to give polymeric $TiCl_3$, a solid with a layer structure and octahedrally coordinated Ti, in finely divided form, that is the catalyst is heterogeneous. Catalysis occurs at crystal defect sites where the metal is coordinatively unsaturated. The surface structure is also thought to be responsible for the stereoregular propene polymerization by imposing steric constraints on the way the monomer can coordinate to the metal prior to insertion into the metal–alkyl bond. It is now commonly accepted that the polymerization follows the principles laid out above, that is the active species is a metal alkyl with a vacant coordination site *cis* to the alkyl ligand (**Cossee–Arlman mechanism**).

The Nobel prize for Chemistry in 1963 was awarded jointly to K. Ziegler and G. Natta.

Metallocene dihalides Cp_2MCl_2 (M = Ti, Zr, Hf) in the presence of aluminium alkyls give highly active *homogeneous* Ziegler–Natta catalysts; the most effective activator is methylaluminoxane (MAO, $[MeAlO]_n$), a polymeric substance with, on average, one methyl group per aluminium atom.

The function of the aluminium alkyl is to:
- alkylate the transition metal;
- act as a Lewis acid and create a vacant coordination site by abstracting a halide or alkyl ligand from the transition metal.
- act as a cleaning role, by mopping up impurities from the monomer and the reaction medium.

Cationic 14 VE alkyl complexes $[Cp_2M-R]^+$ have been identified as the active species. The same cationic complexes are obtained in the absence of aluminum alkyls from metallocene dialkyls by protolysis or R^- abstraction with CPh_3^+. With non-coordinating counterions such as $B(C_6F_5)_4^-$ these cationic catalysts show very high activity. A number of ligand stabilized complexes $[Cp_2MR(L)]^+$ have been isolated (L = THF, RCN, PR_3) which catalyse alkene polymerizations only if dissociation of L is facile.

$$Cp_2MR_2 \quad \xrightarrow{\text{[HNMe}_2\text{Ph]}^+ \text{ X}^- \; / -NMe_2Ph \; ; \; CPh_3^+ / -RCPh_3} \quad [Cp_2M-R]^{\oplus} \; X^{\ominus} \quad M = Ti, Zr, Hf$$

The mechanism of polymer chain propagation is similar to that for heterogeneous catalysts. In either case two coordination sites are required: one for the alkyl group, and one for binding of the unsaturated substrate.

The activity increases in the series M = Ti << Hf < Zr.

Why are these catalysts so highly active? Catalysis only occurs if the metal is in oxidation state IV (d^0); so stabilization of the intermediate olefin adduct by back-bonding is not possible. The d-orbitals of early transition metals are high in energy and, if occupied, would lead to substantial back-bonding stabilization which would greatly increase the activation barrier of the alkyl migration step. For this reason the d^2 complex $Cp_2NbEt(C_2H_4)$, which structurally resembles the intermediate ethene adduct above, resists alkyl migration to the C=C double bond, even under forcing conditions.

Whether or not a metal complex acts as a polymerization, oligomerisation or dimerization catalyst, depends on the likelihood of chain termination. In the absence of hydrogen, alkyl chain growth is usually terminated by β-H elimination (if aluminium alkyls are present, transfer of the polymer chain from the transition metal to aluminium can also take place). Successful polymerization therefore depends on two kinetic parameters: the rate of chain propagation k_p, and the rate of termination k_t.

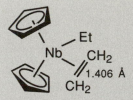

Back-bonding stabilization becomes less important with decreasing *d*-orbital energy from left to right of the periodic table, and coordinatively unsaturated alkyls of cobalt and nickel are indeed able to polymerize ethene, although the activity is substantially lower than with d^0 systems.

If $k_t \approx k_p$, ethene is dimerized to butene (using $NiX_2/AlEt_3$ mixtures). Suitable choice of ligands is important: for example, nickel catalysts can be made selective for the dimerization, oligomerization and polymerization of ethylene. With d^0 complexes $[Cp_2MR]^+$ (M = Ti, Zr, Hf) β-H elimination is not a facile process: $k_p \gg k_t$. Similarly, there are isoelectronic neutral scandium, yttrium and lanthanide complexes Cp'_2MR ($Cp' = C_5H_5$, C_5Me_5) which exhibit high polymerization activity for the same reasons.

Reactions with nucleophiles

Attack of a nucleophile on a metal alkyl often induces reductive coupling.

The reaction is frequently part of a catalytic cycle and leads in effect to the functionalization of alkenes. An example is the hydrocyanation of alkenes.

The hydrocyanation of butadiene with HCN gives adiponitrile, a nylon precursor monomer. It is catalysed by nickel phosphite complexes. While phosphine complexes would react irreversibly to give $Ni(CN)_2L_2$, the more electron-withdrawing phosphite ligands stabilize Ni^0 species even in the presence of HCN. Ethene reacts similarly to propionitrile; the reaction serves to illustrate the principle reaction steps:

A related reaction is the palladium or nickel catalysed cross-coupling of aryl, benzyl, alkenyl or vinyl halides with organometallic reagents, such as zinc, tin, boron, mercury etc. alkyls. The reaction is a frequently used method for the formation of C bonds in organic synthesis:

The Heck arylation of olefins involves the insertion of suitably activated olefins into Pd–C bonds formed by the oxidative addition of aryl halides (usually bromides) to Pd^0. β-H elimination produces the olefinic coupling product:

Interactions of metals with C–H bonds

Whereas alkenes and alkynes have an extensive chemistry and are easily functionalized with a variety of reagents, C–H bonds are generally unreactive, particularly those of saturated hydrocarbons – not surprising in view of the high C–H bond dissociation energies (*c.* 360–450 kJ mol^{-1}). Two processes are possible:

☞ *electrophilic attack on C–H, involving metals in high oxidation states;* and

☞ *oxidative addition of a C–H bond to a coordinatively unsaturated metal centre.*

These may proceed intra- or intermolecularly.

Alkane complex   Hydrido(alkyl) complex

The first step in these reactions is the interaction of a metal centre with a C–H bond or a C–H hydrogen. Interactions of this kind also precede α- and β-hydride elimination processes discussed earlier (see p. 43) where they may be close to the *transition state*. However, in many electron deficient metal systems M•••H–C structures represent the *ground state*. Remarkably compounds such as $[CpCo(L)(CH_2CH_2-\mu-H)]^+$ [L = P(OMe)$_3$, PMe$_3$] are *c.* 20–28 kJ mol^{-1} more stable than the expected classical hydrido isomer, $[CpCo(L)(H)(\eta^2-C_2H_4)]^+$. The term '**agostic**' has been coined to describe this bonding mode of alkyl ligands.

'Agostic' is derived from the Greek αγοστωι, 'to hold onto oneself', an expression used by Homer.

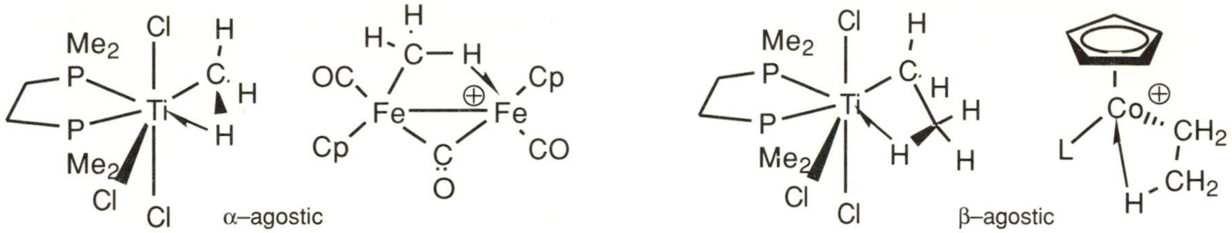

α–agostic   β–agostic

There are numerous reactions in which the 'activation' of a C–H bond, i.e. the substitution of H by a metal, is facile *provided* the moiety carrying the C–H bond remains within the coordination sphere of the metal for a sufficient length of time. In many instances the C–H activation is driven by steric crowding; for example, the alkylation of transition metal salts with such bulky alkyls as CH_2CMe_3 and CH_2SiMe_3, or thermolysis often leads to formation of metallacycles via γ-C–H activation.

Agostic bonding can often be detected by its characteristic NMR parameters, especially the C–H coupling constant (^{13}C NMR). For $[CpCo(L)(CH_2CH_2-\mu-H_a)]^+$: δ^1H for $H_a = -12.6$ ppm, $J_{CH} = 63$ Hz, whereas the other hydrogens are found between 0.2 and 4 ppm, with $J_{CH} \approx 155$ Hz.

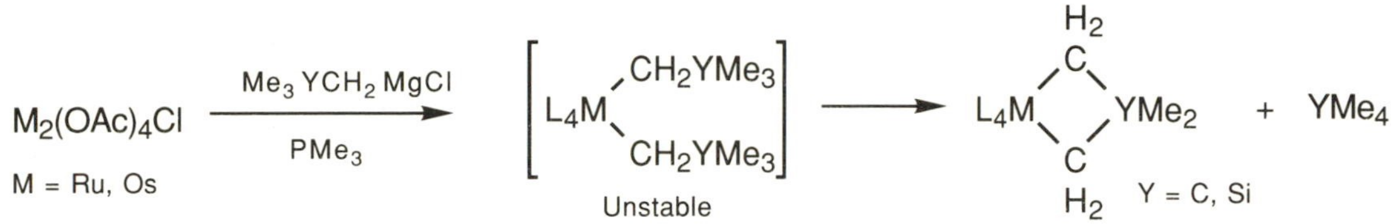

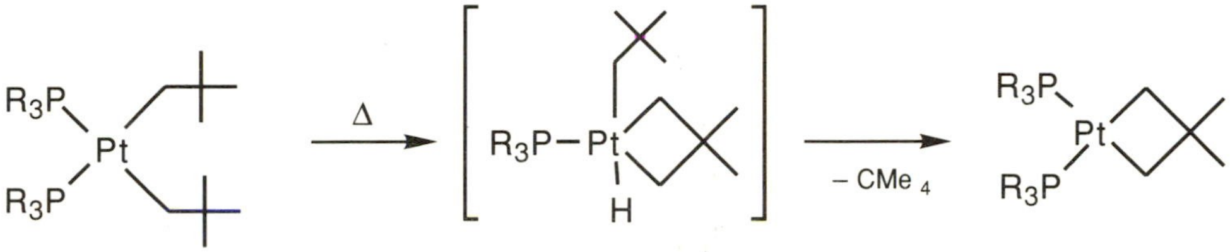

The activation barrier towards *intramolecular C–H activation* is apparently low in many cases. *Intermolecular C–H activation* is, of course, much more attractive, particularly if it can be made catalytic, but for a long time remained an elusive goal.

Before a reaction can occur, a complex between metal and substrate has to be formed, and one reason for the lack on reactivity of saturated hydrocarbons is their extremely poor coordinating ability. There is evidence for the existence of alkane complexes of $Cr(CO)_5$ in matrices at low temperature, and H/D exchange processes are thought to involve a methane complex:

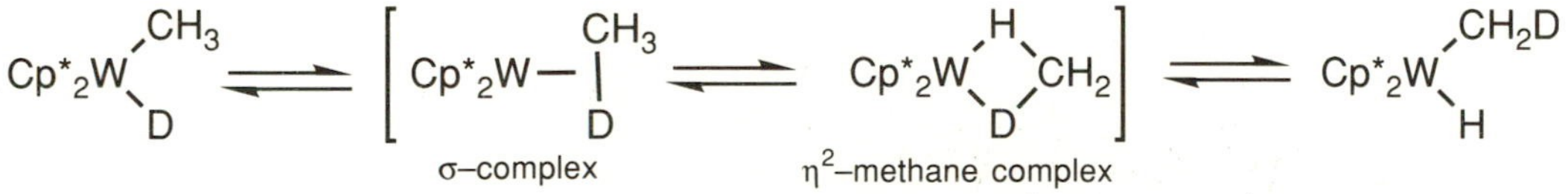

There are no stable alkane complexes, but the coordination mode of methyl compounds might provide a model. Since CH_3^- is isoelectronic to BH_4^- for which numerous hydrogen-bridged systems are known, complexes with one, two or three bonds to alkyl groups ought to be expected. The ytterbium adduct with MeBeCp* comes close to an η^3 alkane complex:

Arenes are better ligands than alkanes and consequently undergo C–H activation much more readily. The first example of intermolecular C–H activation by electron-rich metal fragments was reported by J. Chatt in 1965 who observed that during attempts to reduce $RuCl_2(dmpe)_2$ to $Ru(dmpe)_2$ in the presence of naphthalene a hydrido naphthyl complex was formed (dmpe = $Me_2PCH_2CH_2PMe_2$). In the absence of naphthalene 'Ru(dmpe)$_2$' reacts with the C–H bonds of the phosphine methyl groups

The activation of aromatics is also possible under photolytic conditions and has been shown to proceed via an η^2-arene complex:

Alkane activation can be facilitated if the product is particularly stable and the hydrogen can be transferred to a suitable acceptor. Iridium hydrides react with cyclopentane to form cyclopentadienyl complexes in the presence of $Bu^tCH=CH_2$ (neohexene) as H-acceptor. The reaction proceeds in a series of steps involving successive dehydrogenation of cyclopentane via alkene and allylic intermediates.

The inertness of alkanes such as CH_4 can be overcome if the metal complex intermediate is unstable enough. This is achieved by generating coordinatively unsaturated intermediates of metals which are known to form very stable M–C bonds, such as iridium. Irradiation of $Cp^*Ir(H)_2(PMe_3)$ leads to the dissociation of H_2 to give a coordinatively unsaturated fragment which reacts with alkanes to give (hydrido)(alkyl) complexes. It reacts even with methane under pressure. The product is in many cases surprisingly stable – most alkyl hydride complexes decompose rapidly at low temperature (see p. 43).

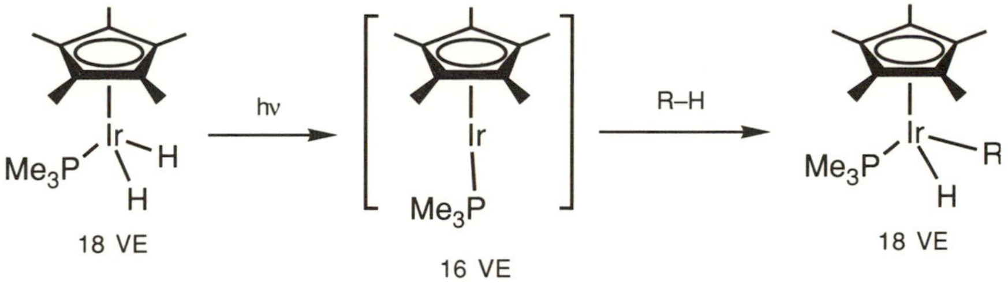

18 VE   16 VE   18 VE

The reactions described so far are stoichiometric. Catalytic alkane activation is achieved with RhCl(CO)(PMe$_3$)$_2$ under photolytic conditions. CO dissociates to give highly reactive RhCl(PMe$_3$)$_2$ which dehydrogenates alkanes; several hundred turnovers per rhodium atom have been observed. Even the direct hydroformylation of alkanes is possible if the reaction is carried out in a CO/H$_2$ atmosphere.

The irradiation of Cp*Ir(CO)$_2$ in fluorocarbon solvents in the presence of R–H similarly gives Cp*Ir(H)(R)(CO) compounds, indicating that electron donation by PMe$_3$ is not an important factor.

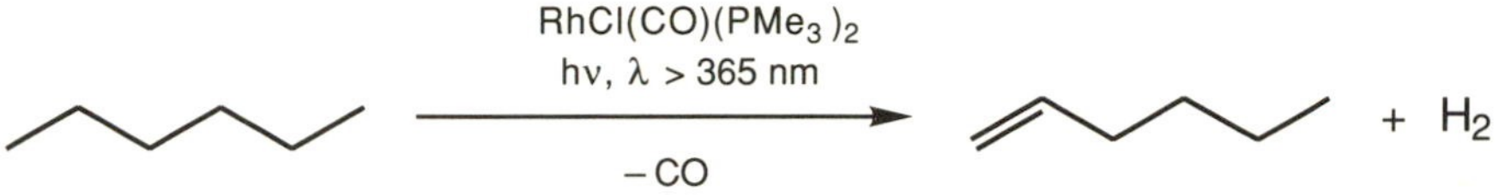

$$\text{RhCl(CO)(PMe}_3)_2 \quad h\nu, \lambda > 365 \text{ nm} \quad - \text{CO}$$

$$+ \; H_2$$

In contrast to the reactions of nucleophilic metal systems described above, the *electrophilic C–H activation* (metallation) of hydrocarbons is well established. The mercuration of benzene has been known since 1898 (O. Dimroth):

PtCl$_4{}^{2-}$ catalyses the H/D exchange between arenes and the solvent D$_2$O/CD$_3$COOD, presumably via H–Pt–Ph intermediates, and the reaction of arenes with H$_2$PtCl$_6$ gives isolable complexes [Ph–PtCl$_4$(NH$_3$)]$^-$. Palladium acetate converts even methane and adamantane into alkylesters, presumably via palladium alkyl intermediates:

$$R\text{–}H + Pd^{2+} \longrightarrow R\text{–}Pd^+ + H^+ \xrightarrow{\text{CF}_3\text{COOH}} R\text{–}OOCCF_3 + Pd^0 + H^+$$

Electrophilic early transition metal complexes reacts with hydrocarbons under exchange of the alkyl ligands. This process has become known as σ-*bond metathesis*. Cp*$_2$ScCH$_3$ exchanges with ^{13}CH$_4$ and gives vinyl, alkynyl and aryl complexes with 1-alkenes, alkynes and arenes, respectively:

$$Cp^*_2Sc-CH_3 \ + \ R-H \ \rightleftharpoons \ \left[Cp^*_2Sc \begin{smallmatrix} R \\ \diagup \diagdown \\ \diagdown \diagup \\ C \end{smallmatrix} H \right] \ \longrightarrow \ Cp^*_2Sc-R \ + \ CH_4$$

$R = {}^{13}CH_3, \ CH=CR'_2, C\equiv CR', \ aryl$

3.5 Transition metal alkyl complexes *in vivo*

Other examples of kinetically inert alkyl complexes include $[R-Cr(H_2O)_5]^{2+}$, $[R-Co(CN)_5]^{3-}$ and $Co(C\equiv CR)_3(NH_3)_3$.

Dorothy Crowfoot–Hodgkin was awarded the Nobel prize in 1964 for her crystallographic work, including the structure determination of vitamin B_{12}.

Whereas most metal alkyl complexes discussed so far do not survive under physiological conditions and many are susceptible to hydrolysis, complexes of d^3 and d^6 ions, Cr^{III} and Co^{III}, are substitutionally inert, the Co compounds in particular. They form alkyl complexes which are stable over a wide pH range.

Cobalt alkyls were for a long time considered rather unstable. It was a great surprise when in 1961 it was discovered by X-ray crystallography that the so–called vitamin B_{12} coenzyme contained a substituted alkyl bound to a cobalt(III) centre. Vitamin B_{12} itself carries a cyanide ligand, an artefact from its isolation using cyanide solution. The metal is bound to a corrin ring and carries a N–base ligand in axial position; the complex is diamagnetic (low spin d^6). Vitamin B_{12} derivatives, the **cobalamines**, are stable to air and aqueous media, and there are numerous alkyl derivatives, such as the methyl complex (methylcobalamine, see below). Vitamin B_{12} catalyses 1,2–alkyl shifts (isomerase reaction):

$$\begin{smallmatrix} R & H \\ | & | \\ -C-C- \\ | & | \end{smallmatrix} \ \underset{}{\overset{B_{12} \text{ coenzyme}}{\rightleftharpoons}} \ \begin{smallmatrix} H & R \\ | & | \\ -C-C- \\ | & | \end{smallmatrix}$$

The Co–C bonds in cobalamines are relatively weak, $c.$ 85–125 kJ mol^{-1}. This means that bond homolysis is fairly facile, and the formation of alkyl radicals is possible, for example induced by light.

Co–C bond cleavage (alkyl transfer) in alkylcobalamines can be induced by three types of reactions: thermolytic or photolytic bond homolysis (radical formation), reactions with nucleophiles, and reactions with electrophiles. Soil bacteria contain methylcobalamin; its reaction with electrophilic Hg^{2+} ions is responsible for the generation of $MeHg^+$, which introduces mercury into the food chain.

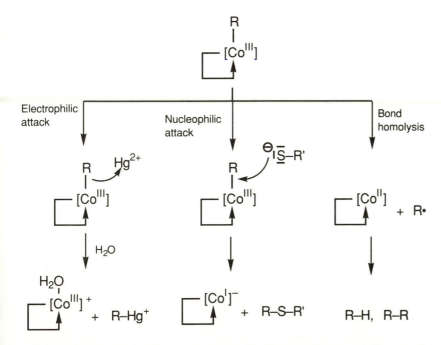

Vitamin B$_{12}$. R in the coenzyme B$_{12}$ is a 5'–deoxyadenosyl group. The corrin ring is indicated by bold lines.

It is a characteristic of these macrocyclic Co complexes that they can be reversibly reduced in a sequence of steps. Hydroxycobalamine, B$_{12b}$, and methylcobalamine are reduced by hydrogen to give a Co^{II} complex, B$_{12r}$.

Further reduction gives the blue–green, extremely air-sensitive B_{12s}, which contains Co^I and reacts both as a metal hydride and as a very strong nucleophile, considerably stronger than, for instance, $SnPh_3^-$ ('supernucleophile').

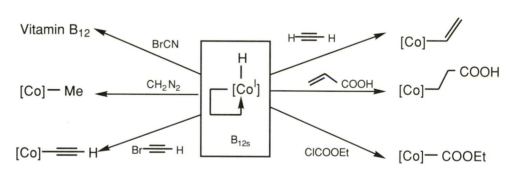

Orange, B_{12} Brown, B_{12r} Blue–green, B_{12s}

Alkylcobalamines are prepared by alkylation of X-cobalamines with Grignard reagents. The importance of radical reactions is illustrated by the high-yield syntheses of alkylcobalamines from vitamin B_{12r} (Co^{II}) with radicals:

The reactions of vitamin B_{12} are closely modelled by simple Co^{III} chelate complexes, such as dimethylglyoximato complexes which have very similar chemical and redox properties.

The reduced form, B_{12s}, reacts like a Co^I hydride and undergoes insertion and oxidative addition reactions. A wide range of functional groups is tolerated, including carboxylic acid functions:

Further reading

J.A. Martinho Simões and J. L. Beauchamp (1990). Transition Metal–Hydrogen and Metal–Carbon Bond Strengths: The Key to Catalysis. *Chem. Rev.*, **90**, 629.

P. J. Davidson, M. F. Lappert and R. Pearce (1976). Metal σ–Hydrocarbyls MR_n. *Chem. Rev.*, **76**, 220.

M. Brookhart, M. L. H. Green and L. L. Wong (1986), Carbon–Hydrogen–Transition Metal Bonds, *Progr. Inorg. Chem.*, **36**, 1.

R.H. Crabtree (1985), The Organometallic Chemistry of Alkanes. *Chem. Rev.*, **85**, 245.

G. Schrauzer (1976). Vitamin B_{12}. *Angew. Chem. Int. Ed. Engl.*, **15**, 417.

4 Alkylidene and alkylidyne complexes

4.1 Complexes with metal–carbon double bonds

Whereas metal alkyl complexes have a long history, the first examples of compounds with metal–carbon double bonds were not discovered until 1964 by E. O. Fischer. These were of the type $(OC)_5M=C(R)(OR')$ (M = Cr, Mo, W), and since they are formally derived from the coordination of a carbene $:CR_2$ to a metal centre, they have become known as **carbene complexes**. The metal is in a low oxidation state, and the bonding of the carbene ligand is reminiscent to that of CO. A second group of M=C compounds, with a highly oxidized metal centre, was subsequently discovered by R. R. Schrock and described as **alkylidene complexes**; this systematic name has now been adopted for all $M=CR_2$ complexes. An example is $Cp_2Ta(CH_3)(CH_2)$.

The formation of M–C multiple bonds is pronounced for metals which are also known to form M–M multiple bonds readily, most notably Ta, Cr, Mo, W and Re. Mononuclear alkylidene complexes of elements of the iron, cobalt, and nickel triads are much less common. For titanium, zirconium, and hafnium, complexes with M=C double bonds are the exception, and there are no M≡C triply bonded species.

Synthesis of alkylidene complexes in low oxidation states

From metal carbonyls

This is the most widely applied preparative method. Isocyanide complexes react similarly, even with weaker nucleophiles such as secondary amines or alcohols.

An unusual variant of this reaction of is the reduction of coordinated CO with reactive zirconium(II) complexes. Note that zirconium acts here as both the alkylating agent and the Lewis acid. This routes gives access to carbene complexes of a variety of metals.

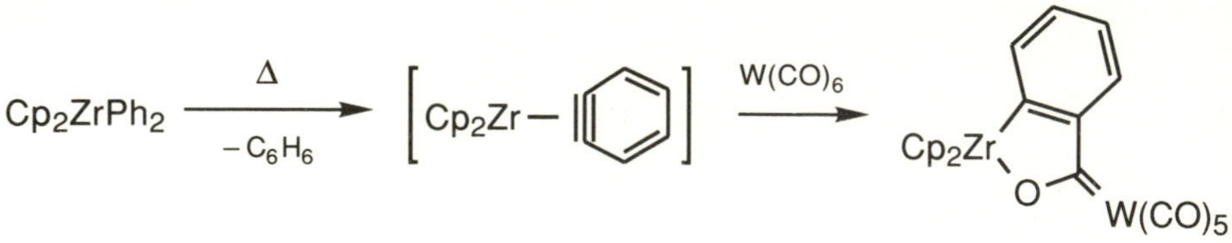

$$Cp_2ZrPh_2 \xrightarrow[-C_6H_6]{\Delta} \left[Cp_2Zr- \parallel\!\bigcirc \right] \xrightarrow{W(CO)_6} Cp_2Zr\cdots O \diagdown C{=}W(CO)_5$$

By rearrangements of coordinated ligands

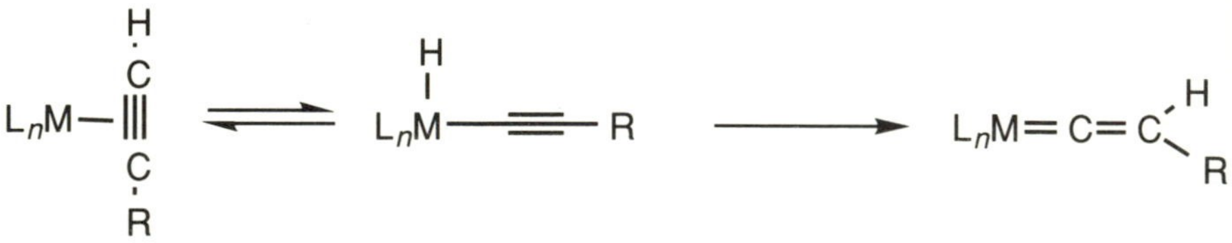

$$L_nM-\overset{\overset{\displaystyle\cdot H}{\underset{\displaystyle|}{C}}}{\underset{\underset{\displaystyle R}{\displaystyle\cdot C}}{|||}} \rightleftharpoons L_nM\overset{\overset{\displaystyle H}{|}}{=\!\!\equiv\!\!=}R \longrightarrow L_nM{=}C{=}C\overset{H}{\underset{R}{\diagdown}}$$

$L_nM = CpMn(CO)_2$, $CpRe(CO)_2$, $CpML_2^+$ (M = Fe, Ru, Os), CpRhL

From activated olefins

Very electron-rich olefins, such as tetrakis(dialkylamino) substituted ethylene, are able to react with noble metals under C=C bond cleavage. Ru and Pt complexes of this type are known.

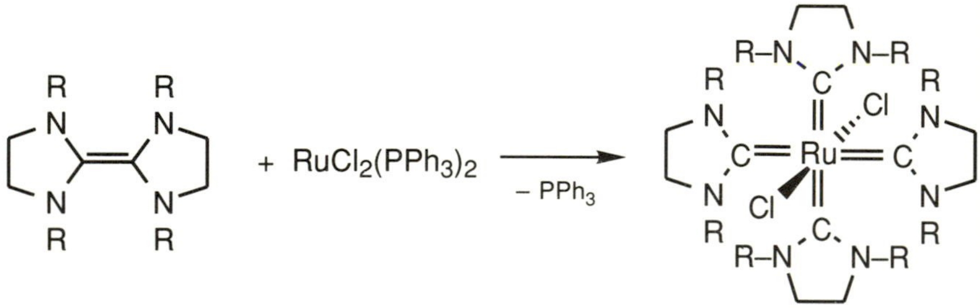

Strained cyclic olefins can be ring-opened to give alkylidene compopunds:

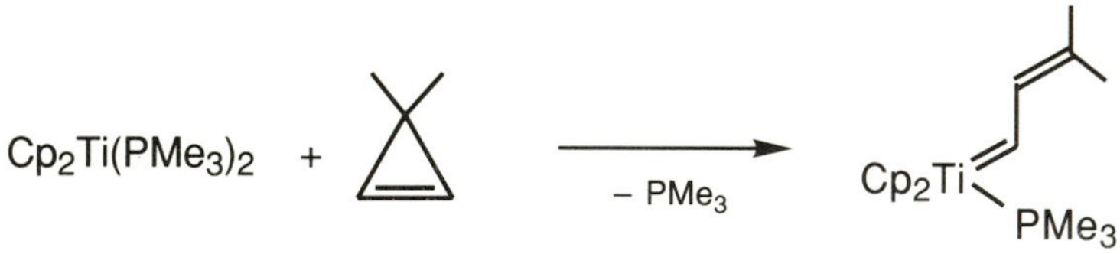

$$Cp_2Ti(PMe_3)_2 + \triangle \!\!\!\bigtriangleup \xrightarrow{-PMe_3} Cp_2Ti{=}\cdots$$

From carbene precursors

$$R\cdots M\text{—Cl} \quad \xrightarrow[\;-\,N_2,\;]{\text{CH}_2\text{N}_2} \quad$$

M = Ru, O

By reaction with electrophiles

$$\text{Cp*L}_2\text{Fe–CH}_2\text{OMe} \quad \xrightarrow[\;-\;\text{MeOH}\;]{\text{H}^+} \quad \left[\,\text{Cp*L}_2\text{Fe}=\!\!=\text{CH}_2\,\right]^+$$

Cp* = C_5Me_5

L = phosphine

The reaction of metal carbonyl anions with iminium salts gives access to heteroatom–free carbene complexes:

$$\text{W(CO)}_5{}^{2-} + [\text{Me}_2\text{N=CPh}_2]^+ \longrightarrow \left[(\text{OC})_5\text{W}\!\!\underset{\text{Ph}}{\overset{\text{Ph}}{\diagdown}}\!\!\text{NMe}_2\right]^- \xrightarrow[\;-\;\text{Me}_2\text{NH}_2{}^+\;]{\text{H}^+} (\text{OC})_5\text{W}=\text{C}\underset{\text{Ph}}{\overset{\text{Ph}}{\diagup}}$$

Synthesis of alkylidene complexes in high oxidation states

By α-hydrogen abstraction

Whereas the alkylation of $TaCl_5$ with MeLi gives $TaMe_5$, it is not possible to accommodate five alkyl ligands if these are substantially more bulky. Whereas electron-rich noble metals in similar sterically congested situations tend to form metallacycles via γ-C–H activation (see Chapter 3), early transition metal complexes such as tantalum undergo *α-H abstraction* to give alkylidene complexes. α-H abstraction is also induced if alkyl halide complexes are treated with strong donor ligands such as PMe_3.

$$(\text{Bu}^t\text{CH}_2)_3\text{TaCl}_2 \quad \xrightarrow{\;\text{Bu}^t\text{CH}_2\text{Li}\;} \quad (\text{Bu}^t\text{CH}_2)_3\text{Ta}=\text{C}\underset{\text{Bu}^t}{\overset{\text{H}}{\diagup}}$$

$$\text{Bu}^t\text{CH}_2\text{–TaCl}_4 \quad \xrightarrow[\;\text{PMe}_3\;]{\;\text{Na/Hg}\;} \quad (\text{PMe}_3)_3(\text{H})\text{Cl}_2\text{Ta}=\text{C}\underset{\text{Bu}^t}{\overset{\text{H}}{\diagup}}$$

With tantalum this process is surprisingly facile. The reaction of $[Cp_2TaMe_2]^+$ with sodium methoxide might have been expected to lead to the addition of the nucleophile to the metal; instead, a methyl ligand is deprotonated to give the first example of a methylene complex in high yield:

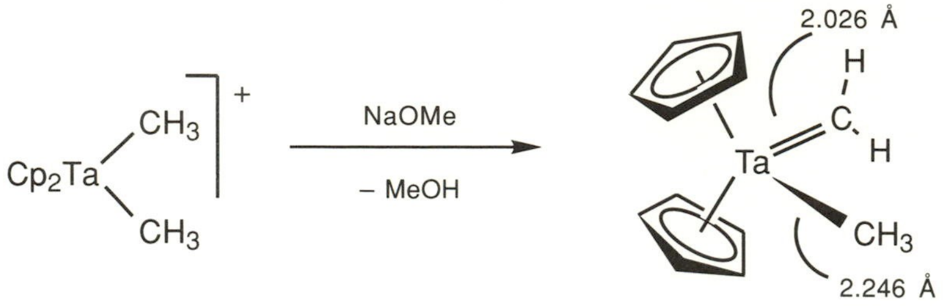

Intramolecular α-C–H activation is reversible. There is evidence that some methyl complexes are in equilibrium with the corresponding hydrido-methylene isomers; these may be trapped by the addition of donor ligands:

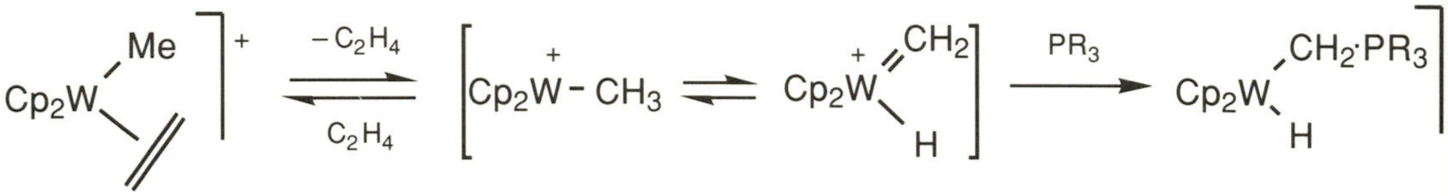

Bonding in alkylidene complexes

The bonding of the carbene ligand in low valent metal complexes ('Fischer carbenes') can be described by resonance hybrid structures which indicate the role of the heteroatom substituent as electron donor. The carbene carbon is electrophilic, as shown in structure **A** which makes an important bonding contribution.

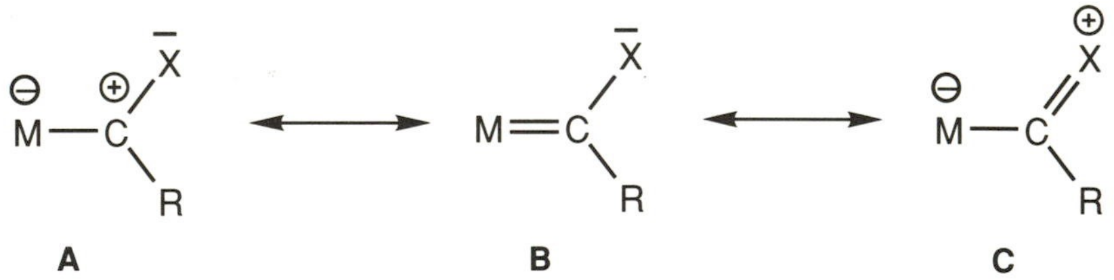

Most of these complexes are stabilized by heteroatom substituents on carbon.

By contrast, and rather contrary to 'intuition', the alkylidene carbon in high oxidation state complexes behaves as a nucleophile. Evidently, in this case the metal acts as a two-electron donor, whereas in carbonyl complexes it behaves as a two-electron acceptor. Heteroatoms such as O or N, with their non-bonding electron pairs, would not provide any stabilization, and indeed are absent in these complexes.

These differences help to explain on the one hand the reaction of Fischer carbenes with lithium alkyls, and on the other, the formation of adducts between alkylidenes with Lewis acids, such as aluminium alkyls:

Electrophilic carbene complex

Nucleophilic alkylidene complex

The electron deficient character of high valent metal centres often leads to strong M–C–H (α–agostic) interactions, with remarkably narrow M–C–H angles (<90°).

$(OC)_5W = C \big\langle \begin{smallmatrix} OMe \\ Ph \end{smallmatrix}$ $\xrightarrow{\text{PhLi}}$ $\left[(OC)_5W - C \big\langle \begin{smallmatrix} OMe \\ -Ph \\ Ph \end{smallmatrix} \right]^-$ $\xrightarrow[-78\ °C,\ -\ MeOH]{\text{HCl}}$ $(OC)_5W = C \big\langle \begin{smallmatrix} Ph \\ Ph \end{smallmatrix}$

In some cases the isomeric (alkylidyne)(hydrido) complexes also exist.

ν_{C-H} 2200 cm^{-1} !

T–shaped CH$_2$ ligand

(Alkylidyne)(hydrido) complex

P–P = dmpe

L = PMe$_3$

4.2 Complexes with metal–carbon triple bonds

After the discovery of compounds with metal–carbon double bonds, metal–carbon triply bonded systems provided a logical synthetic challenge. As often happens, the route to such complexes was found accidentally, in an attempt to prepare halogen-substituted carbene complexes.

As with carbene complexes, there are M≡C–R stabilized by carbonyls and related ligands, and complexes of metals in high oxidation states (d^0) such as Ta^V, containing halides and other anionic ligands. The assignment of the metal oxidation state obviously depends on whether carbyne is considered as a neutral three–electron donor or as a trianion.

The first metal carbyne complex was discovered by E. O. Fischer, in 1973.

Synthesis

From low-valent metal complexes

The reaction of 'Fischer carbene' complexes with boron or aluminum halides leads to the abstraction of the heteroatom substituent on carbon. The M–C–C unit is generally linear; the M–C distance is shorter than bonds to carbenes or to CO.

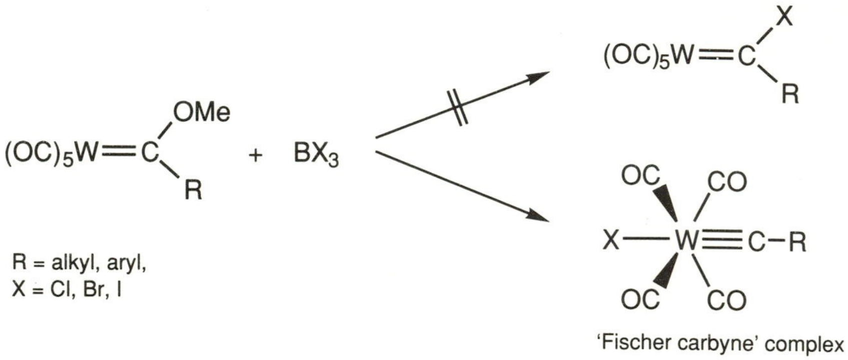

$$(OC)_5W=C\underset{R}{\overset{OMe}{<}} \quad + \quad BX_3$$

R = alkyl, aryl,
X = Cl, Br, I

$$(OC)_5W=C\underset{R}{\overset{X}{<}}$$

$$X-W\equiv C-R$$

'Fischer carbyne' complex

Acid halides may be employed as Lewis acids in the reaction with acyl anions:

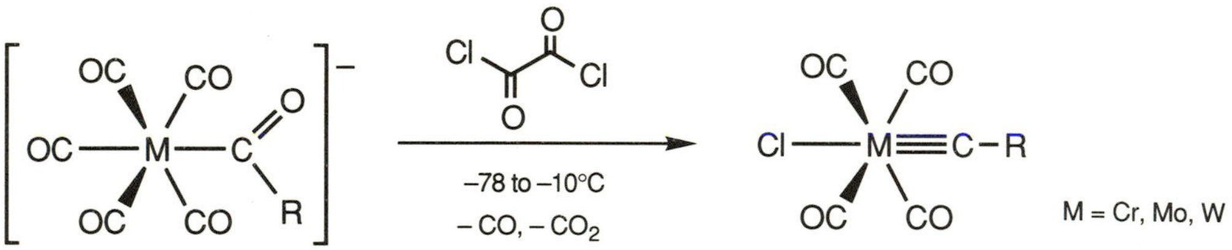

$$\left[\ OC-M(CO)_4-C\underset{R}{\overset{O}{<}}\ \right]^- \xrightarrow[-CO,\,-CO_2]{\substack{-78\ to\ -10°C}} Cl-M\equiv C-R$$

M = Cr, Mo, W

From alkylidene complexes

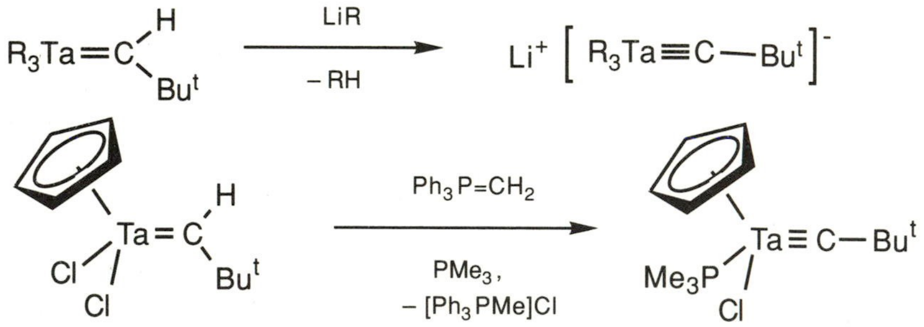

$$R_3Ta=C\underset{Bu^t}{\overset{H}{<}} \xrightarrow[-RH]{LiR} Li^+\left[\ R_3Ta\equiv C-Bu^t\ \right]^-$$

$$Cp(Cl)_2Ta=C\underset{Bu^t}{\overset{H}{<}} \xrightarrow[\substack{PMe_3,\\ -[Ph_3PMe]Cl}]{Ph_3P=CH_2} Cp(Me_3P)(Cl)Ta\equiv C-Bu^t$$

Other routes

M≡C triple bonds, particularly to tungsten, are sometimes formed in a number of unexpected ways, such as during alkylations, or from alkynes by C≡C bond cleavage.

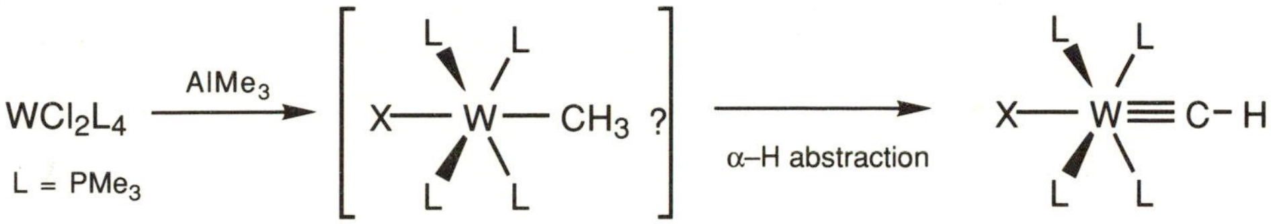

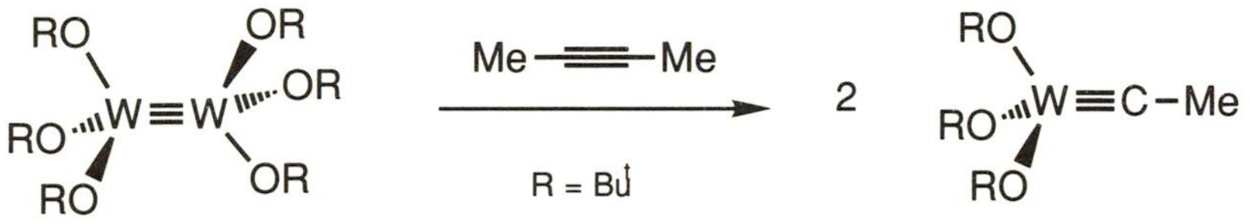

Dichlorocarbenes are highly reactive. Treatment with aryl lithium leads to double halide abstraction:

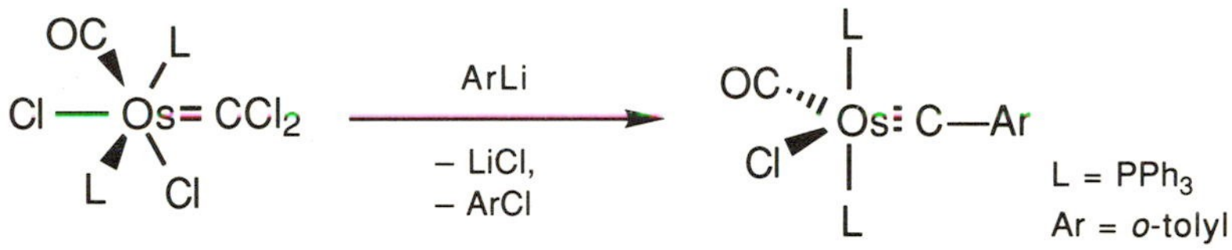

Bonding in alkylidyne complexes

The M≡C–R unit has acetylene-like character, with two π systems perpendicular to each other. The M–C–C skeleton is close to linear (*sp*-hybridized carbon).

There are complexes which contain both alkyl and alkylidyne ligands, and even examples where all three types of hydrocarbyl ligands are bound to the same metal. Although one could envisage a process whereby the α-hydrogens of an alkyl ligand might migrate to and from alkylidene and/or alkylidyne α-carbons, there does not appear to be a facile pathway, and each ligand retains its distinct identity.

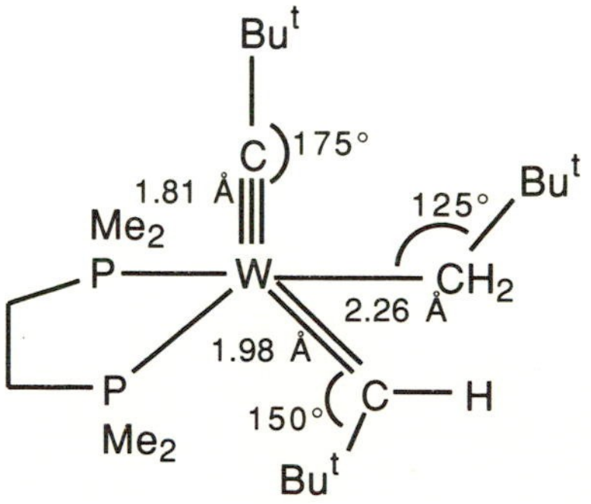

4.3 Reactivity of alkylidene and alkylidyne complexes

Reactions with electrophiles

Alkylidene complexes $LnM=CR_2$ have a formal similarity with ketones $O=CR_2$ and with phosphorus ylids, $R'_3P=CR_2$ (Wittig reagents) and indeed similar chemical behaviour has been observed. Alkylidene complexes free of heteroatoms and ligands such as CO are the most reactive; due to the oxophilic character of early transition metals they readily react with carbonyl compounds under carbene transfer to give olefins:

$$R_3Ta=C\begin{smallmatrix}H\\ \\Bu^t\end{smallmatrix} + O=C\begin{smallmatrix}R\\ \\R'\end{smallmatrix} \longrightarrow [R_3TaO]_n + \begin{smallmatrix}H\\Bu^t\end{smallmatrix}{=}\begin{smallmatrix}R\\R'\end{smallmatrix}$$

$R = Bu^tCH_2$

Alkylidenes and alkylidynes react readily with Brønsted acids. This is a common method for the interconversion of alkylidynes and alkylidenes, for the transformation into alkyls, and of vinylidene complexes into alkylidynes.

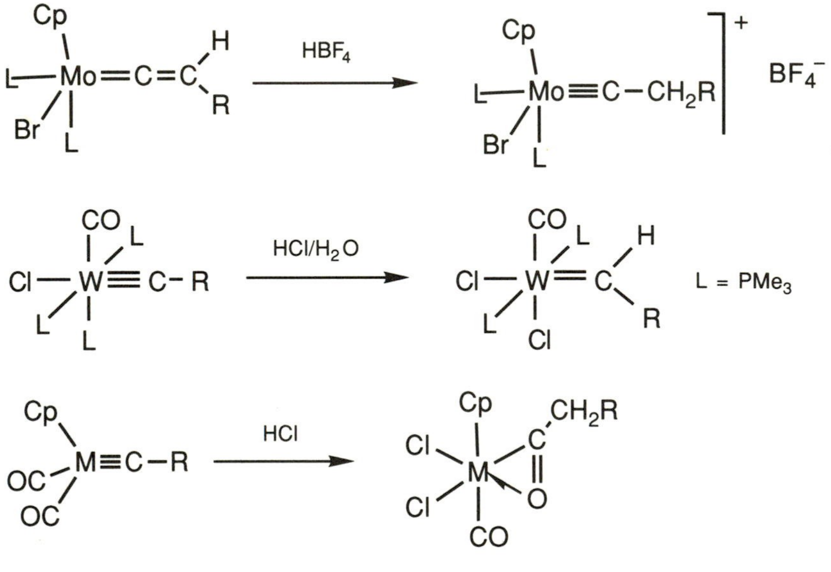

Alkylidene transfer

Alkylidene ligands have been shown to undergo intermolecular exchange reactions:

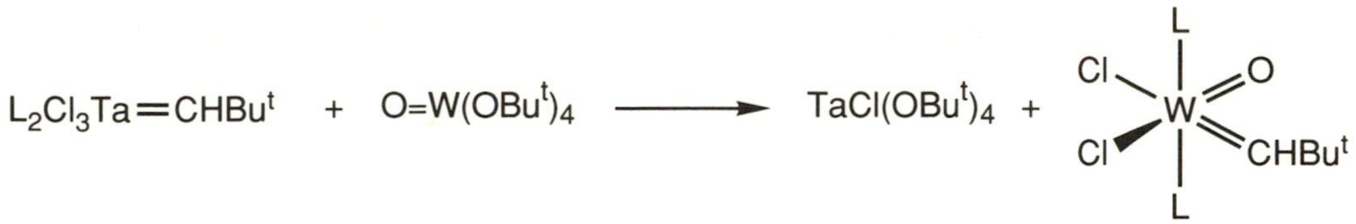

$$L_2Cl_3Ta=CHBu^t \ + \ O=W(OBu^t)_4 \ \longrightarrow \ TaCl(OBu^t)_4 \ +$$

(structure: octahedral W complex with L (axial), =O, CHBu^t, and two Cl, L (axial))

Alkylidene–alkyl coupling reactions

Alkylidene ligands may react with nucleophiles either intra- or intermolecularly with formation of new C–C bonds. Reactions of this kind have been discussed as models for the formation of hydrocarbons from unstable surface species during the reduction of CO (Fischer–Tropsch reaction, see p. 33).

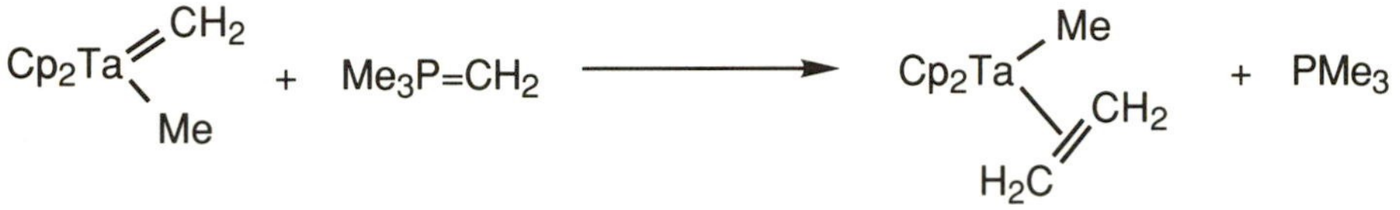

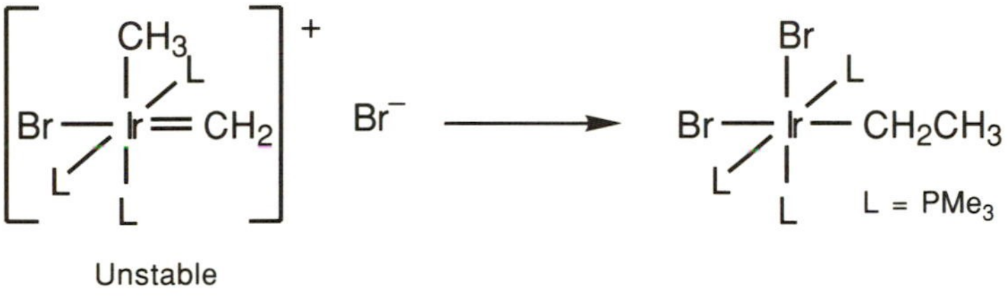

$$Cp_2Ta(=CH_2)(Me) \ + \ Me_3P=CH_2 \ \longrightarrow \ Cp_2Ta(Me)(CH_2=CH_2 \ portion) \ + \ PMe_3$$

$$[Br\text{--}Ir(CH_3)(L)_2(L)=CH_2]^+ \ Br^- \ \longrightarrow \ Br\text{--}Ir(Br)(L)_2(L)\text{--}CH_2CH_3 \quad L = PMe_3$$

Unstable

Reactions with alkenes and alkynes

The reactivity of alkylidenes with unsaturated organic molecules is widely exploited in synthesis. The insertion and cyclization reactions with alkynes are typical examples.

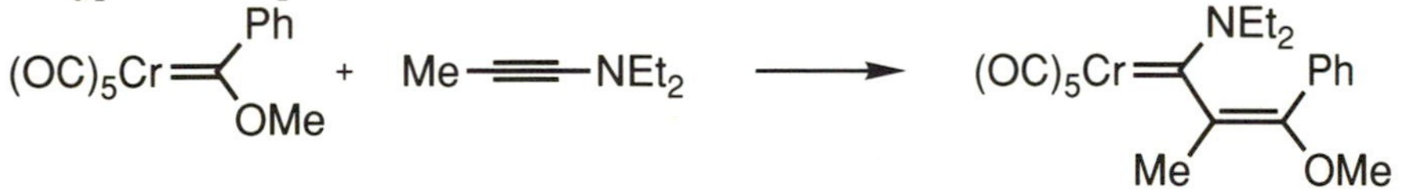

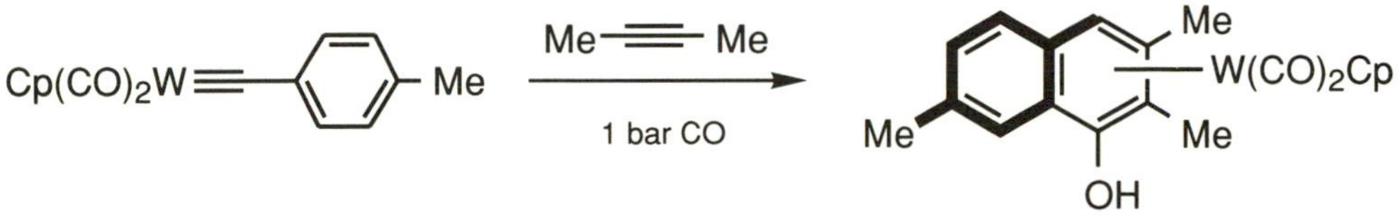

$$(OC)_5Cr=C(Ph)(OMe) \ + \ Me\!\equiv\!NEt_2 \ \longrightarrow \ (OC)_5Cr=C(NEt_2)\text{-}C(Ph)=C(Me)(OMe)$$

$$Cp(CO)_2W\equiv\!\!-\!\!\langle\text{C}_6\text{H}_4\rangle\!\!-\!\!Me \ \xrightarrow[\text{1 bar CO}]{Me\equiv Me} \ \text{(naphthol complex)}$$

A naphthol complex, formed from alkylidyne-alkyne coupling. The original alkylidyne moiety is indicated by bold lines.

A compound with particular widespread synthetic applications is the **Tebbe reagent** which can be regarded as the protected form of the unstable ('$Cp_2Ti=CH_2$'). Many of its reactions are complementary to Wittig reagents.

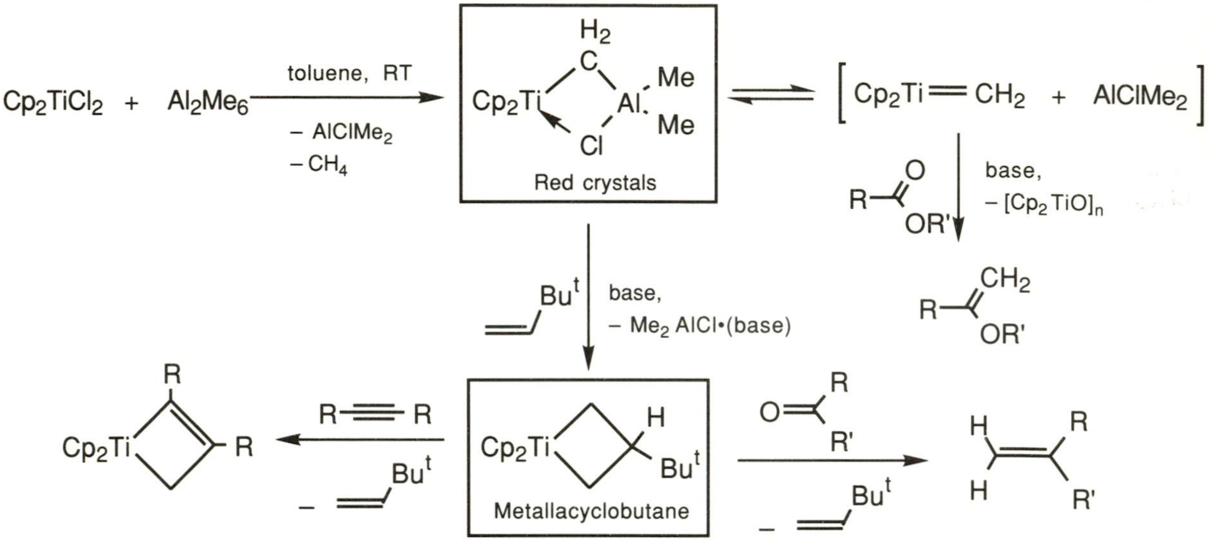

A characteristic reaction is the ability of alkylidenes and alkylidynes to form metallacycles. This tendency is exploited in organic synthesis, in olefin metathesis, and for the formation of alkylidene- and alkylidyne-bridged heteronuclear metal clusters.

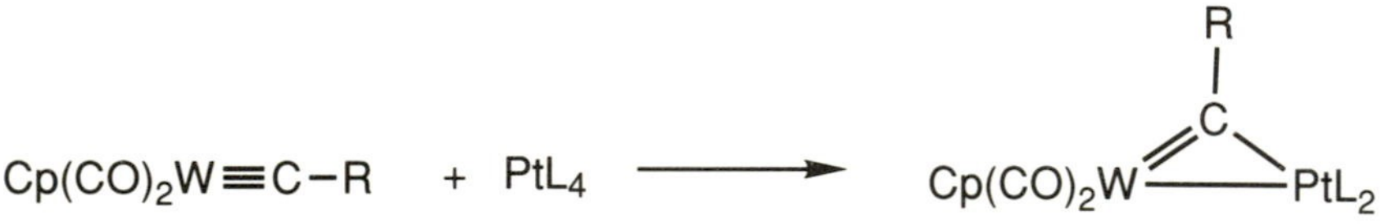

Olefin metathesis

The development of metal alkylidene chemistry was of fundamental importance for the elucidation of a rather unusual reaction, the **olefin metathesis** (*H. Eleuterio*):

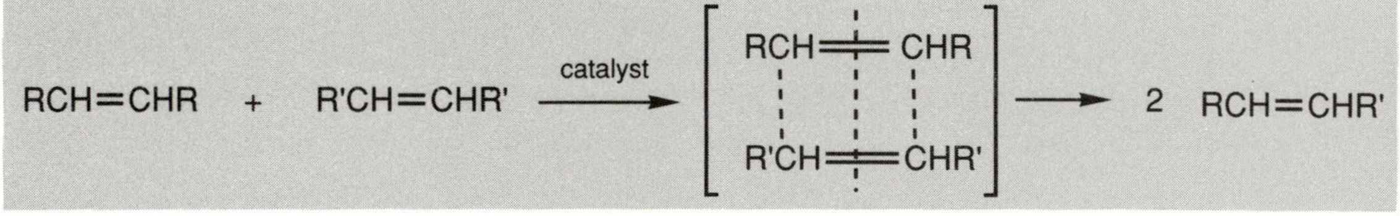

Conventional catalysts are based on MoCl₅ or WCl₆/SnMe₄ mixtures, or on rhenium oxides. The catalytically active species are now known to be alkylidene complexes, and several well-defined complexes are available which are highly active either in the presence or absence of Lewis acids.

The olefin metathesis process is a consequence of the tendency of alkylidene complexes to form metallacycles with unsaturated substrates. As with the Tebbe reagent, the metallacycle formation is reversible and generates the new olefin together with a new alkylidene species.

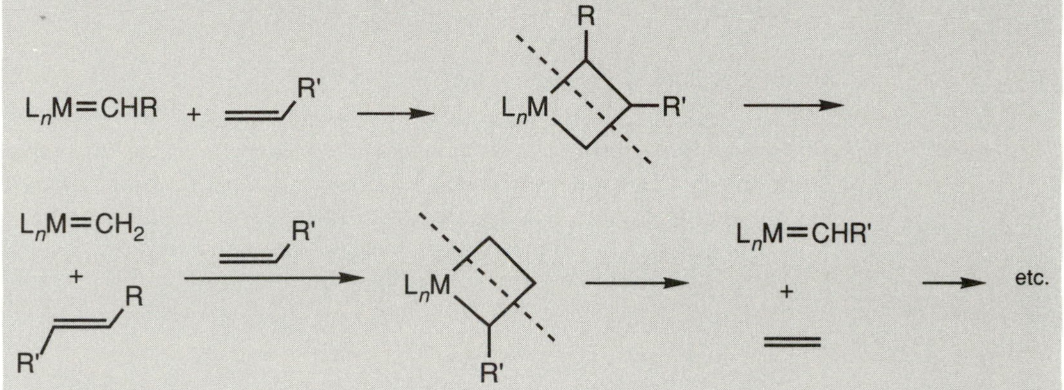

With well-defined alkylidene catalyst precursors, the consumption of the initial complex, together with the appearance of chain carrying alkylidene species and metallacycles has been demonstrated by NMR spectroscopy.

With cyclic olefins this process leads to **ring opening metathesis polymerization (ROMP):**

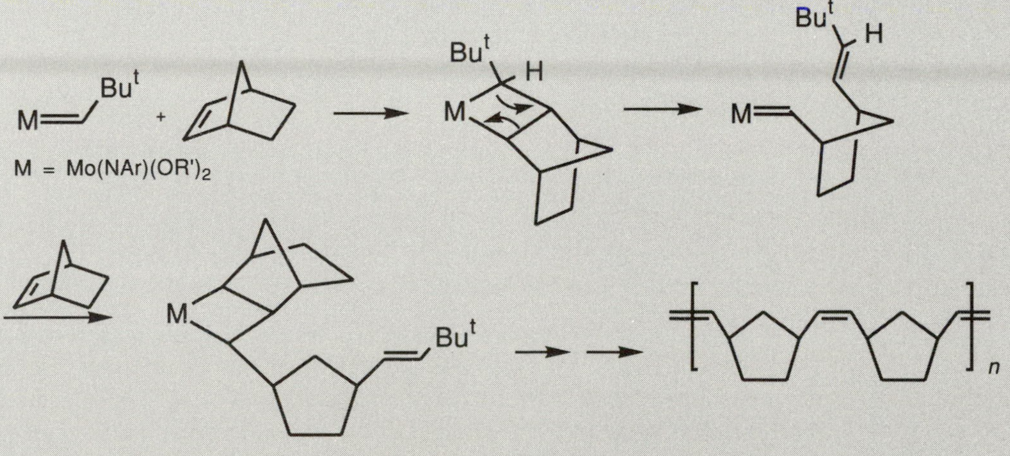

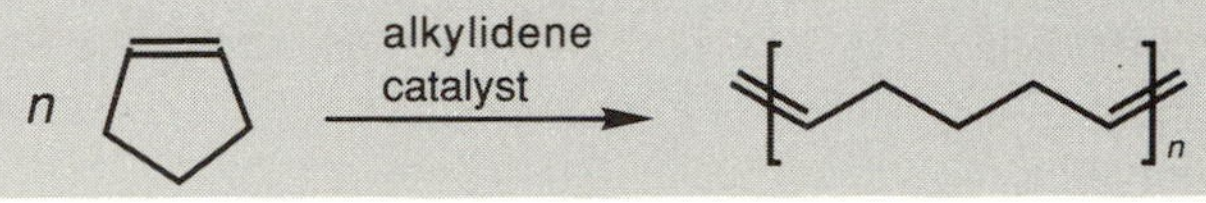

Molybdenum alkylidene complexes have proved to be particularly active single-component ROMP catalysts. They are also active for the metathesis of functionalized olefins such as unsaturated esters. Ruthenium(II), particularly $[Ru(H_2O)_6]^{2+}$, catalyses the ring-opening metathesis of norbornene derivatives even in water, and although little is known about the mechanism in this case, it is likely to proceed via alkylidenes also. A ruthenium vinylidene complex has been isolated which catalyses the ROMP of norbornene in alcoholic media.

Alkyne metathesis

Alkynes are metathesized in a way similar to alkenes, via metallacyclic intermediates which can be isolated in some instances. Terminal alkynes undergo metathesis less readily; they form metallacyclobutadienes whose hydrogen-bearing carbon is readily deprotonated. Metallacyclobutadienes form delocalized π-systems.

4.4 Dinuclear alkylidene and alkylidyne complexes

There are numerous homo- and heterodinuclear complexes with bridging alkylidene and alkylidyne ligands. Complexes of this kind have made an important contribution towards the understanding of C–C bond forming processes during the hydrogenation of CO (Fischer–Tropsch reaction, see p. 33). Alkyl, alkylidene and alkylidyne species are often interconvertible by the successive additions of H^+ and H^-. The reactions are exemplified by the ruthenium complexes shown below; there are similar iron, cobalt, and rhodium compounds.

Apart from these 18 VE complexes, α-H abstraction during the alkylation of metal salts with Grignard or lithium reagents occasionally leads directly to

CR$_2$ and CR complexes, such as Ru$_2$(μ_2-CH$_2$)$_3$(PMe$_3$)$_6$ and M$_2$(μ_2-CSiMe$_3$)$_2$(CH$_2$SiMe$_3$)$_6$ (M = Nb, W, Re).

Further reading

H. Fischer, F.R. Kreissl, U. Schubert, P. Hofmann, K. H. Dötz, and K. Weiss, (1984). Transition Metal Carbene Complexes. VCH ; Weinheim.

R. R. Schrock (1986), High-Oxidation-State Molybdenum and Tungsten Alkylidyne Complexes, *Acc. Chem. Res.*, **19**, 342.

A. Mayr and H. Hoffmeister (1991). Recent Advances in the Chemistry of Metal–Carbon Triple Bonds. *Adv. Organomet. Chem*, **92**, 227.

For information on any aspect of M–C, M=C and M≡C chemistry see:

G. Wilkinson, F. G. A. Stone and E. W. Abel (eds.) (1982). *Comprehensive Organometallic Chemistry.* Pergamon, Oxford.

Index